7th Grade
Rhode Island Math
for Beginners

Standardized Testing
and Home Schooling
Study Guide

MATHFA

7th Grade Rhode Island Math for Beginners

By Mathfa, Email: Support@mathfa.com

Web: www.mathnotion.com

Copyright © 2025 by Mathfa, an imprint of Math Notion Inc. All rights reserved. No part of this publication may be reproduced, stored in a retrieval system, or transmitted in any form or by any means, electronic, mechanical, photocopying, recording, scanning, or otherwise, except as permitted under Section 107 or 108 of the 1976 United States Copyright Act, without permission of the author.

All inquiries should be addressed to Mathfa.

ISBN: 978-1-63620-369-0

Welcome to Math Grade 7!

Seventh grade is a turning point where familiar ideas like fractions and decimals team up with new friends such as linear equations and probability. This book invites you to travel into that exciting middle ground with confidence and curiosity. Inside, chapters unfold like mini-quests: you'll tackle integer arithmetic, explore proportional thinking, investigate geometry, and dig into data, all while uncovering how these concepts connect to everyday life.

You'll find clear explanations followed by carefully chosen examples that light the path forward. Guided practice lets you test your skills right away, while optional "Think Deeper" problems encourage you to look beyond the obvious and spot hidden patterns.

Helpful hints appear exactly when they're needed, so you always have a gentle boost nearby. Progressively challenging problems build solid confidence before you climb to the next concept, and quick reflection prompts invite you to pause, celebrate what you know, and set intentions for the lessons ahead.

Throughout these pages you'll meet real-world scenarios that show math as a tool for creativity and problem-solving, not just a list of rules. The book is designed for everyone: independent learners, homeschooling families seeking structure, classroom teachers crafting engaging lessons, and parents eager to support their children's progress. By the time you finish, you'll not only understand math, but you'll also trust yourself to tackle any numbers that come your way.

Most important, this guide aims to nurture a growth mindset. Mistakes become steppingstones, questions spark discovery, and every small breakthrough builds momentum for the next. Let this book be both map and companion as you transform today's challenges into tomorrow's achievements. Turn the page and let your Grade 7 math adventure begin.

www.mathnotion.com

www.mathnotion.com

… So Much More Online!

✓ FREE Mathematics Worksheets

✓ More Math Learning Books!

✓ FREE Math Lessons

✓ Online Math Tutors

✓ **PDF Version of This Book**

No Registration Required!

Visit us here!

How to effectively use the Study Guide?

Seventh-grade math is where the foundations you built earlier start linking together into bigger ideas; fractions meet algebra, numbers meet geometry, and problems solving takes center stage. This guide is written to make that journey clear and motivating, whether you're preparing for your state test or learning independently at home. Follow the strategies below to turn each page into real progress:

1) **First, Preview the Roadmap**

    Spend a few minutes with the table of contents and chapter openers. Seeing the full journey ahead makes every stop along the way feel purposeful.

2) **Set Personal Milestones**

    Decide what you want from each unit, mastering integer operations, nailing two-step equations, or boosting overall confidence. Clear targets keep your effort focused.

3) **Craft a Study Rhythm—at Your Speed**

    Block out regular study pockets that fit your week. Whether you learn best in short daily bursts or longer weekend sessions, keep it steady and adjust the pace to match your comfort level.

4) **Work Through Examples, Not Just Around Them**

    Before reading the solution, cover it up and try the example yourself. Compare your steps, mark differences, and rewrite any part that felt shaky.

5) **Dive Into the Practice Sets**

    Every skill gets stronger through use. Complete the guided exercises right after the lesson, then tackle the independent problems to prove you've got it.

6) **Pause to Fix Mistakes**

    When an answer is off, circle it and jot a quick note on what tripped you up, maybe a sign error, a rushed calculation, or a misunderstanding term. This reflection turns errors into insights.

7) **Stretch with "Challenge Corners"**

    Each chapter ends with tougher questions designed to push your reasoning. Treat them like a mini-boss battle: optional but rewarding.

www.mathnotion.com

8) Loop Back Weekly

Reserve one session a week to revisit earlier topics. A quick rework of a few past problems keeps older skills fresh while you learn new ones.

9) Keep a Progress Log

List chapters finished, scores on self-quizzes, and concepts that still feel unclear. Watching the checklist fill up is great motivation.

10) Spot Math in Real Life

Convert a recipe, compare sale prices, or break down sports stats. Real-world practice shows why these skills matter beyond the page.

11) Built for School Standards

Every lesson follows the key ideas schools and statewide expects you to know, so you're getting ready for any big test without even noticing.

12) Ask, Share, Repeat

Stuck? Talk it through—with a friend, parent, tutor, or online forum. Explaining a question aloud often reveals the answer, and hearing someone else's approach can spark a fresh insight.

This book moves with you. Explore, practice, question, and celebrate small wins; the confidence you build today will shape a stronger, bolder math mind for tomorrow!

Contents

Chapter 1: Integers and Number Theory 11
    Integers and Additive Inverse 12
    Adding and Subtracting Integers 13
    Multiplying and Dividing Integers 14
    Order of Operations 15
    Operation With Absolute Value 16
    Ordering Integers 17
    Prime and Composite Numbers 18
    Factoring Numbers 19
    Greatest Common Factor (GCF) 20
    Least Common Multiple (LCM) 21
    Word Problems 22
    Worksheets 23
    Answer of Worksheets 27

Chapter 2: Decimals 29
    Comparing Decimals 30
    Adding and Subtracting Decimals 31
    Multiplying and Dividing Decimals 32
    Multiplying Decimals by Power of Ten 33
    Dividing Decimals by Power of Ten 33
    Rounding Decimals 34
    Word Problems 35
    Worksheets 36
    Answer of Worksheets 38

Chapter 3: Fractions 39
    Simplifying Fractions 40
    Improper Fractions 41
    Least Common Denominator 42
    Converting to Decimals 43
    Converting to Fractions or Mixed Numbers 43
    Ordering Mixed Numbers and Improper Fractions 44
    Adding and Subtracting 45
    Multiplying and Dividing 46
    Mixed Operations 47
    Complex Fraction Operations 48
    Word Problems 49
    Worksheets 50
    Answer of Worksheets 53

Chapter 4: Rational Numbers 55
    Rational Numbers 56
    Ordering Rational Numbers 57

Adding and Subtracting Rational numbers 58
Multiplying and Dividing Rational Numbers 59
Mixed Operation on Rational Numbers 60
Absolute Value Operation 61
Word Problems 62
Worksheets 63
Answer of Worksheets 65

Chapter 5: Proportions, Ratios and Percent 67

Simplifying Ratios 68
Understanding Proportions 69
Solving Proportions 70
Similarity and Ratios 71
Ratios and Rates Word Problems 72
Percentage Calculations 73
Discount, Tax and Tip 74
Simple Interest 75
Compound Interest 76
Word Problems 77
Worksheets 78
Answer of Worksheets 82

Chapter 6: Exponents and Radical Expressions 83

Multiplication Property of Exponents 84
Division Property of Exponents 85
Negative Exponents and Negative Bases 86
Fraction and Decimal Bases 87
Scientific Notation 88
Square Roots 89
Estimate Square Roots 90
Word Problems 91
Worksheets 92
Answer of Worksheets 94

Chapter 7: Algebraic Expressions 95

Translating Phrases into an Algebraic Statement 96
Identify Terms, Coefficients, and Simplifying 97
Properties of Addition and Multiplication 98
Evaluating Variable Expressions 99
Factor by Distributive Property 100
Factor by Area Model 101
Word Problems 102
Worksheets 103
Answer of Worksheets 105

Chapter 8: Equations and Inequalities 107

One-Step Equations 108
Multi-Step Equations 109
One-Step Inequalities 110
Multi-Step Inequalities 111
Graphing Inequalities 112

  Word Problems 113
  Worksheets 114
  Answer of Worksheets 116

Chapter 9: Linear Functions 118
  Slope from Graph 119
  Slope from Two Points 119
  Identifying Slope and Y-Intercept 120
  Rate of Change 121
  Graphing Lines Using Line Equation 122
  Writing Linear Equations 123
  Write an Equation from Graph 124
  Graphing Linear Inequalities 125
  System of Equations 126
  Point Coordinate and Finding Midpoint 127
  Finding Distance of Two Points 128
  Word Problems 129
  Worksheets 130
  Answer of worksheets 134

Chapter 10: Transformations 137
  Translations 138
  Reflections 139
  Rotations 140
  Dilations 141
  Scale Drawings 142
  Worksheets 143
  Answer of Worksheets 145

Chapter 11: Sequences 146
  Identify Arithmetic Sequences 147
  Identify Geometric Sequences 148
  Evaluate Variable Expressions 149
  Word Problems 150
  Worksheets 151
  Answer of Worksheets 153

Chapter 12: Congruence and Similarity 154
  Similar Figures 155
  Ratio of Area and Volume in Similar Figures 156
  Similarity Criteria for Triangles 157
  Similar Figures and Indirect Measurement 158
  Congruent Figures 159
  Congruence Criteria of Triangles 160
  Word Problems 161
  Worksheets 162
  Answer of Worksheets 166

Chapter 13: Geometry and Solid Figures 169
  Angles of Triangles and Quadrilateral 170
  Interior and Exterior Angles of Polygons 171

Angle of Circle 172
Pythagorean Theorem 173
Area of Compound Figures 174
Perimeter of Compound Shapes 175
Front, Side and Top of Three-Dimensional Figures 176
Cubes 177
Rectangular Prism 178
Cylinder 179
Pyramids 180
Cones 181
Word Problems 182
Worksheets 183
Answer of Worksheets 189

Chapter 14: Statistics and Probability 191

Mean and Median 192
Mode and Range 193
Mean Absolute Deviation 194
Frequency Chart 195
Histograms 196
Pie Graph 197
Quartiles and Outlier 198
Box and Whisker Plot 199
Probability of Simple Events 200
Probability of Opposite Events 201
Probability of Mutually Events 202
Theoretical Probability 203
Experimental Probability 204
Make Predictions 205
Compound Events 206
Probability Word Problems 207
Worksheets 208
Answer of Worksheets 215

Chapter 15: Practice Test Review 219

Reference Material 220
RICAS Practice Test 221
Answer Key 229
Answers and Explanations 231

7th Grade Rhode Island Math

Chapter 1: Integers and Number Theory

Topics that you'll learn in this chapter:

- ✓ Integers and Opposite Integers (Additive Inverse)
- ✓ Adding and Subtracting Integers
- ✓ Multiplying and Dividing Integers
- ✓ Order of Operations
- ✓ Operation With Absolute Value
- ✓ Ordering Integers
- ✓ Prime and Composite Numbers
- ✓ Factoring Numbers
- ✓ Greatest Common Factor (GCF)
- ✓ Least Common Multiple (LCM)
- ✓ Word Problems
- ✓ Worksheets
- ✓ Answer of Worksheets

www.mathnotion.com

Integers and Additive Inverse

Integers are the set of whole numbers that include all positive numbers, all negative numbers, and zero. They do not include any fractional or decimal numbers.

$$\mathbb{Z} = \{\dots, -3, -2, -1, 0, 1, 2, 3, \dots\}$$

Visual Representation:

Imagine a number line with zero in the middle. Positive integers are to the right of zero, while negative integers are to the left:

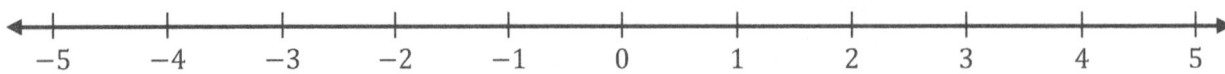

Opposite integers are pairs of numbers that are the same distance from zero on the number line but in opposite directions.

☑ The term "Additive Inverse" is another name for the "Opposite of a number".

Key Points

1. **Zero**: Zero is its own opposite.
2. **Sum**: The sum of an integer and its opposite is always zero (e.g., $-8 + 8 = 0$).

Examples:

1) Find the numbers a, b, c, d and e:

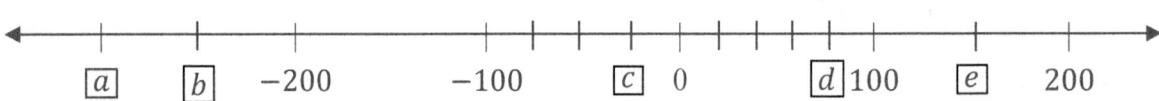

Solution:

a: -300

b: This number is between -300 and -200 so it is -250

c: The interval between -100 and 0 is divided into 4 parts and each part is equal - 25 ($-100 \div 4 = -25$), so c is -25.

d: The interval between 0 and 100 is divided into 5 parts and each part is equal 20 ($100 \div 5 = 20$), so d is 80 ($20 \times 4 = 80$).

e: This number is between 100 and 200 so it is 150.

2) Find integer numbers from the numbers:

$$18, \quad -10000, \quad 0.12, \quad \frac{2}{3}, \quad -(-9), \quad 15.000, \quad \frac{10}{5}, \quad 39.99$$

Solution:

Integer numbers do not include any fractional or decimal numbers, so $18, -10000, -(-9)$ (it is equal to 9), 15.000 (it is equal to 15) and $\frac{10}{5}$ (after simplifying this fraction, it is equal to 2) are integers.

Adding and Subtracting Integers

Adding Integers

1. **Same Signs**: When you add two integers with the same sign, add their absolute values and keep the common sign.
2. **Different Signs:** When you add two integers with different signs, subtract the smaller absolute value from the larger absolute value and keep the sign of the integer with the larger absolute value.

Subtracting Integers

1. **Subtraction as Addition:** Subtracting an integer is the same as adding its opposite.
2. **Using Opposites:** Change the subtraction problem to an addition problem by using the opposite of the integer you're subtracting.

Example:

Find the result of the expressions:

a) $(+7) + (+2)$
b) $-3 + (-5)$
c) $(+6) + (-1)$
d) $(-10) + (+4)$
e) $9 - 13$
f) $-8 - 12$
g) $-2 - (-7)$

Solution:

a) Both numbers are positive, so add the absolute values: $7 + 2 = 9$
   The result is positive: $+9$
b) Both numbers are negative, so add the absolute values: $3 + 5 = 8$
   The result is negative: -8
c) 6 is larger and positive. Subtract the 1 from 6: $6 - 1 = 5$
   The result shows the sign of the number with the larger absolute value: $+5$
d) 10 is larger and negative. Subtract the smaller absolute value from the larger absolute value: $10 - 4 = 6$
   The result shows the sign of the number with the larger absolute value: -6
e) Rewrite the subtraction as addition: $9 + (-13)$
   Follow the rules for adding integers with different signs: $9 + (-13) = -4$
f) Rewrite the subtraction as addition: $-8 + (-12)$
   Follow the rules for adding integers with the same sign: $-8 + (-12) = -20$
g) Rewrite the subtraction as addition: $-2 + (+7)$
   Follow the rules for adding integers with different signs: $-2 + (+7) = +5$

www.mathnotion.com

Multiplying and Dividing Integers

Multiplying Integers: When multiplying integers, the sign of the product depends on the signs of the factors being multiplied.

1. **Same Sign**: The product of two integers with the same sign is positive.
2. **Different Signs**: The product of two integers with different signs is negative.

Dividing Integers: When dividing integers, the sign of the quotient depends on the signs of the dividend and the divisor.

1. **Same Sign**: The quotient of two integers with the same sign is positive
2. **Different Signs**: The quotient of two integers with different signs is negative.

Rule Summary:

- $(+) \times (+) = (+)$
- $(-) \times (-) = (+)$
- $(+) \times (-) = (-)$
- $(-) \times (+) = (-)$

- $(+) \div (+) = (+)$
- $(-) \div (-) = (+)$
- $(+) \div (-) = (-)$
- $(-) \div (+) = (-)$

Example:

Solve.

a) $(+3) \times (+4)$
b) $(-5) \times (-6)$
c) $(+7) \times (-2)$
d) $(-8) \times (+3)$

e) $(+24) \div (+6)$
f) $(-42) \div (-3)$
g) $(+18) \div (-2)$
h) $(-32) \div (+8)$

Solution:

a) The factors have the same signs, so the product is positive: $(+3) \times (+4) = +12$
b) The factors have the same signs, so the product is positive: $(-5) \times (-6) = +30$
c) The factors have different signs, so the product is negative: $(+7) \times (-2) = -14$
d) The factors have different signs, so the product is negative: $(-8) \times (+3) = -24$
e) Both numbers have the same signs, so the quotient is positive: $(+24) \div (+6) = +4$
f) Both numbers have the same signs, so the quotient is positive: $(-42) \div (-3) = +14$
g) Each number has different signs, so the quotient is negative: $(+18) \div (-2) = -9$
h) Each number has different signs, so the quotient is negative: $(-32) \div (+8) = -4$

www.mathnotion.com

Order of Operations

The order of operations is a set of rules that tells you the correct sequence to evaluate a math expression, including those involving integer numbers. It is essential for ensuring consistency and accuracy in solving mathematical problems. The acronym **PEMDAS** helps remember the order of operations:

PEMDAS:

- **P: Parentheses**: Solve expressions inside parentheses or brackets first.
- **E: Exponents**: Solve exponents (or powers and roots) next.
- **MD: Multiplication** and **Division**: Perform multiplication and division from left to right.
- **AS: Addition** and **Subtraction**: Perform addition and subtraction from left to right.

Examples:

1) Evaluate the expression $3 + 6 \times (5 + 4) \div 3^2 - 7$.

   Solution:

   1. Parentheses (the addition inside the parentheses): $3 + 6 \times (9) \div 3^2 - 7$
   2. Exponents: $3 + 6 \times (9) \div 9 - 7$
   3. Multiplication and Division (left to right):

      $3 + 54 \div 9 - 7$
      $3 + 6 - 7$

   4. Addition and Subtraction (left to right):
      $3 + 6 - 7 = 9 - 7 = 2$

   So, the final result is 2.

2) Evaluate the expression $3 - [(-25 \div 5 + 2 \times 6) + 2^5 \div (-8)]$.

   Solution:

   1. Innermost parentheses:
      - Division and multiplication inside the parentheses: $3 - [(-5 + 12) + 2^5 \div (-8)]$
      - Addition inside the parentheses: $3 - [7 + 2^5 \div (-8)]$
   2. Second bracket:
      - The exponent inside the bracket: $3 - [7 + 32 \div (-8)]$
      - The division inside the bracket: $3 - [7 - 4]$
      - The subtraction inside the bracket: $3 - 3$
   3. Subtraction: $3 - 3 = 0$

Operation With Absolute Value

The absolute value of an integer is its distance from zero on the number line, regardless of direction. It is always a non-negative number.

Symbol: The absolute value of an integer a is denoted by $|a|$.

Properties:
1. $|a| \geq 0$
2. $|a| = a$ if $a \geq 0$
3. $|a| = -a$ if $a < 0$
4. $|0| = 0$

Addition and Subtraction: When dealing with absolute values in addition and subtraction, treat them as positive values, then apply the rules for addition and subtraction of integers.

☑ You should note that $|a + b| \neq |a| + |b|$ and $|a - b| \neq |a| - |b|$.

Multiplication and Division: The absolute value of a product or quotient is the product or quotient of the absolute values.

Examples:

1) Find the absolute value of the integer numbers:

   $12, -55, 0, -(-2)$

   Solution:
   - Absolute value of 12: $|12| = 12$
   - Absolute value of -55: $|-55| = 55$
   - Absolute value of 0: $|0| = 0$
   - Absolute value of $-(-2)$: $|-(-2)| = |2| = 2$

2) Do the expressions:

   a) $|5| + |-3|$
   b) $|8| - |-10|$
   c) $|-4 \times (-6)|$
   d) $|12 \div (-2)|$

   Solution:
   a) $|5| + |-3| = 5 + 3 = 8$
   b) $|8| - |-10| = 8 - 10 = -2$
   c) $|-4 \times (-6)| = |-4| \times |-6| = 4 \times 6 = 24$
   d) $|12 \div (-2)| = |12| \div |-2| = 12 \div 2 = 6$

Ordering Integers

Ordering integers involves arranging them from the smallest to the largest (ascending order) or from the largest to the smallest (descending order).

Steps to Order Integers

1. **Identify the Integers**: Write down the integers you need to order.

2. **Compare Their Values**: Determine the value of each integer, considering that negative numbers are smaller than positive numbers.

3. **Arrange from Smallest to Largest**: For ascending order, start with the most negative number and proceed to the largest positive number. For descending order, start with the largest positive number and proceed to the most negative number.

Examples:

1) Order the integers $-7, |-9|, 5, -|3|, |0|$ from greatest to least:
   Solution:
   - $|-9| = 9$
   - $-|3| = -3$
   - $|0| = 0$

   List the result: $9 > 5 > 0 > -3 > -7$

2) Order each set of the following integers:
   a) $-8^2, 0, -(-6), -|-17+8|, -15-3$ (from the smallest to the largest)
   b) $11 - |-11|, -\frac{14}{7}, |-(-3)|, -(12 \div (-6)), -7 \times (1-3)$ (from the largest to the smallest)

   Solution:

   a) Calculate each expression:
   - $-8^2 = -64$
   - $-(-6) = +6$
   - $-|-17+8| = -|-9| = -9$
   - $-15 - 3 = -18$

   List the result: $-64 < -18 < 0 < 6 < 9$

   b) Calculate each expression:
   - $11 - |-11| = 11 - 11 = 0$
   - $-\frac{14}{7} = -2$
   - $|-(-3)| = 3$
   - $-(12 \div (-6)) = -(-2) = 2$
   - $-7 \times (1-3) = -7 \times -2 = +14$

   List the result: $14 > 3 > 2 > 0 > -2$

www.mathnotion.com

Prime and Composite Numbers

Prime Numbers:

A prime number is a whole number greater than 1 that has exactly two distinct positive divisors: 1 and itself. This means a prime number can only be divided evenly (without a remainder) by 1 and by the number itself, such as 2, 3, 5, 7, 11, etc.

Composite Numbers

A composite number is a whole number greater than 1 that has more than two distinct positive divisors. This means a composite number can be divided evenly by 1, itself, and at least one other number, such as 4, 10, 15, 24, etc.

How to Determine if a Number is Prime or Composite

1. **Check for Divisors**: See if the number has any divisors other than 1 and itself.
2. **Prime**: If the only divisors are 1 and the number itself, it's prime.
    - ☑ The other general way is to check divisibility by all prime numbers less than its square root.
3. **Composite**: If it has other divisors, it's composite.

Examples

1) Is the number 51 prime or composite?
   Solution:
   - Check the divisors: 1, 3, 17 and 51.
   - Since there are more than two divisors, 91 is a composite number.

2) Is the number 193 prime or composite?

Solution:

For 193, we check divisibility by all prime numbers less than its square root (approximately 13.89). The prime numbers to check are 2, 3, 5, 7, 11, and 13.

- 193 is not divisible by 2 (it's an odd number).
- The sum of the digits ($1 + 9 + 3 = 13$) is not divisible by 3.
- 193 does not end in 0 or 5, so it's not divisible by 5.
- Dividing 193 by 7 gives approximately 27.57, which is not an integer.
- Dividing 193 by 11 gives approximately 17.55, which is not an integer.
- Dividing 193 by 13 gives approximately 14.85, which is not an integer.

Since 193 is not divisible by any of these prime numbers, it is confirmed to be a prime number.

www.mathnotion.com

Factoring Numbers

Factoring numbers involve breaking down a number into its smaller components, called factors, that can be multiplied together to get the original number. For example, the factors of 12 are 1, 2, 3, 4, 6 and 12 because:

$$1 \times 12 = 12, \quad 2 \times 6 = 12, \quad 3 \times 4 = 12$$

Prime factorization: Breaking down a number into prime numbers that multiply to the original number. For example, $12 = 2^2 \times 3$

Steps to Factor Numbers:

1. **Start with the Smallest Prime Number (2):** Divide the number by 2 if it's even.

2. **Continue with Larger Prime Numbers**: If the number is odd or no longer divisible by 2, try the next smallest prime number (3, 5, 7, etc.).

3. **Repeat Until the Result is 1**: Continue factoring until you reach 1.

Examples:

1) Factor the number 48.
   Solution:
   1. Divide by 2 (since 48 is even):
      $48 \div 2 = 24$
   2. Continue dividing by 2:
      $24 \div 2 = 12$
      $12 \div 2 = 6$
      $6 \div 2 = 3$
   3. 3 is a prime number:
      $3 \div 3 = 1$
   4. Prime Factorization of 48: $48 = 2^4 \times 3$

2) What are the number of distinct prime factors of 150:
   Solution:
   1. Divided by 2 (since 150 is even):
      $150 \div 2 = 75$
   2. Continue dividing by 3 (because $7 + 5 = 12$ and 12 is divisible by 3:
      $75 \div 3 = 25$
   3. Continue dividing by 5 (because its last digit is 5):
      $25 \div 5 = 5$
   4. 5 is a prime number:
      $5 \div 5 = 1$
   5. Prime Factorization of 150: $150 = 5^2 \times 3 \times 2$. So, the number of distinct prime factors is 3 (2, 3 and 5).

Greatest Common Factor (GCF)

The greatest common factor (GCF), also known as the greatest common divisor (GCD), of two or more integers is the largest positive integer that divides all of the given numbers without leaving a remainder.

How to Find the GCF (Prime Factorization Method):

1. Find the prime factorization of each number.
2. Identify the common prime factors.
3. Multiply the smallest power of all common prime factors.

Examples:

1) Find the greatest common factor of 84 and 72.
   Solution:
   1. Find the prime factorization of 84 and 72:
      - Prime factorization of $84 = 2^2 \times 3 \times 7$
      - Prime factorization of $72 = 2^3 \times 3^2$
   2. Identify the common prime factors: The common prime factors are 2 and 3
   3. Multiply the smallest power of all common prime factors: For 2, the lowest power is 2^2 and for 3 the lowest power is 3^1, so the GCF is $2^2 \times 3 = 12$.

2) Find the greatest common factor of $-36, -54$ and 90.
   Solution:
   To find the Greatest Common Factor (GCF) of $-36, -54$, and 90, we can ignore the negative signs and focus on the positive values 36, 54, and 90. The GCF will be the same for both their positive and negative values:
   1. Find the prime factorization of 36, 54 and 90:
      - Prime factorization of $36 = 2^2 \times 3^2$
      - Prime factorization of $54 = 2 \times 3^3$
      - Prime factorization of $90 = 2 \times 3^2 \times 5$
   2. Identify the common prime factors: The common prime factors are 2 and 3
   3. Multiply the smallest power of all common prime factors: For 2, the lowest power is 2^1 and for 3 the lowest power is 3^2, so the GCF is $2^1 \times 3^2 = 18$.

www.mathnotion.com

Least Common Multiple (LCM)

The least common multiple (LCM) is the smallest positive integer that is divisible by each of the given numbers.

How to Find the LCM (Prime Factorization Method):

1. Find the prime factorization of each number.
2. Identify all the prime factors that appear in any of the factorizations.
3. For each prime factor, take the highest power that appears in any of the factorizations.
4. Multiply these highest powers together to get the LCM.

Examples:

1) Find the least common multiple of 48 and 75.

   Solution:

   1. Find the prime factorization of 48 and 75:
      - Prime factorization of 48: $48 = 3 \times 2^4$
      - Prime factorization of 75: $75 = 3 \times 5^2$
   2. Identify all the prime factors that appear in factorizations: 3, 2 and 5
   3. Take the highest power that appears in any of the factorizations:
      - For 2, the highest power is 2^4. For 3, the highest power is 3^1 and for 5, the highest power is 5^2.
   4. Multiply these highest powers together to get the LCM: $LCM = 2^4 \times 3 \times 5^2 = 1,200$.

2) For the two following numbers the $GCF = 3^2 \times 7$ and $LCM = 2^3 \times 3^2 \times 5^2 \times 7$, find a, b, c, d, e and f:

   $A = 2^a \times 3^b \times 7^c$ and $B = 3^d \times 5^e \times 7^f$

   Solution: The common prime factors are 3 and 7. The GCF indicates min (b, d) =2 and min (c, f) =1. The LCM indicates:
   - For 2: $a = 3$ (Since A is the only number with 2 as a factor)
   - For 3: $b = d = 2$
   - For 5: $e = 2$ (Since B is the only number with 5 as a factor)
   - For 7: $c = f = 1$

   So, $a = 3, b = 2, c = 1, d = 2, e = 2$ and $f = 1$.

Word Problems

Solving word problems involving integer numbers and number theory requires a systematic approach:

Steps to Solve Integer and Number Theory Word Problems:

1. **Understand the Problem**:
   - Read the problem carefully.
   - Identify what is being asked.
   - Highlight or underline key information and numbers.
2. **Identify Known and Unknown Values**:
   - Determine the known values given in the problem.
   - Identify the unknowns you need to find.
3. **Translate the Problem into Mathematical Expressions**:
   - Convert the word problem into algebraic expressions or equations.
   - Use variables to represent unknowns.
4. **Solve the Equations**:
   - Use appropriate mathematical operations to solve the unknowns.
   - Follow the order of operations (PEMDAS) where necessary.
5. **Check Your Solution**:
   - Substitute your solution back into the original problem to verify if it makes sense.
   - Ensure the solution is reasonable and answer the question being asked.

Example:

A number is added to -15, and the result is multiplied by -3 to give 45. Find the number.

Solution:

1. Understand the Problem: Identify that we need to find the number that, when added to -15 and multiplied by -3, gives 45.

2. Identify Known and Unknown Values: Let the unknown number be x. The equation based on the problem: $(x + (-15)) \times (-3) = 45$

3. Translate the Problem: Simplify the equation : $(x - 15) \times (-3) = 45$

4. Solve the Equation: Divide both sides by -3:

   $x - 15 = -15$ so $x = 0$

5. Check Your Solution: Substitute $x = 0$ back into the original problem: $(0 - 15) \times (-3)$

   Simplify: $-15 \times (-3) = 45$

   The solution is correct, and the number is 0.

Worksheets

Integers and Additive Inverse

Find missing integer numbers (1, 2, 3 and 4):

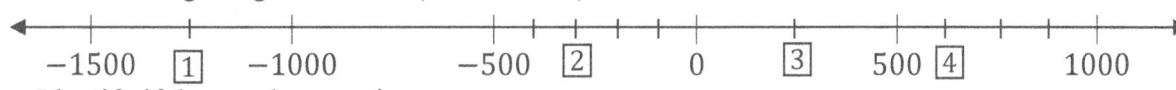

Identify if the numbers are integers or not:

5) $-185/0002$
6) $\frac{15}{3}$
7) -1264
8) π

Write the opposite of each number relative to -2:

9) 7
10) -3

Adding and Subtracting Integers

Do the following additions and subtractions:

1) $-8 + 4$
2) $7 - (-2)$
3) $-10 + 15 - 5$
4) $12 - 8 + (-4)$
5) $-20 + 5 - 15 + 10$
6) $25 - (-5) + 8 - (-3)$
7) $(-25 + 14) - (-17) + (-9)$
8) $[-(8 + (-5) - (-4))] - (-2 + 3)$
9) $[(-10) + (6 - 4) - 8] + [15 - 2]$
10) $(-5) + 3 - [(-7) + 12] - (6 - 2)$

Multiplying and Dividing Integers

Do the following multiplications and divisions:

1) -5×6
2) $-18 \div 3$
3) $6 \times (-4) \times 2$
4) $(-15) \div 3 \times (-2)$
5) $(-7) \times (-8) \div (-4)$
6) $(5 \times (-3) \times 2) \div (-6)$
7) $(-12 \div 4) \times (-3) \div 9 \times (-2)$
8) $[(-4) \times 6 \div (-3)] \times [5 \div (-1) \times 2]$
9) $[(-9) \times (-2) \div 3] \times [(8 \div (-4)) \times 5]$
10) $-6 \times [(-4) \times (-5)] \div 2 \times (-3) \times (-8)$

Order of Operations

Evaluate each expression:

1) $3 + 4 \times 2$
2) $(5 + 3) \times 2$
3) $12 \div (3 + 1) \times 2$
4) $8 - (3 \times 2 - 4) \div 2$
5) $6 + [2 \times (3 - 8) + 4] \div 2$
6) $3^2 \times (4 - 2) + 5^2$
7) $(2 + 3)^2 - 4 \div 2 + 2^2$
8) $7 \times [(2^2 + 3^1) \div (5 - (-2))]$
9) $(-1^2 \times 4^2) - (4 \div (-2))^2$
10) $-(-[12 \div (-4) + (-5 - 6)] + (-9 - 1)^2)$

Operation With Absolute Value

Evaluate.

1) $| 3 - 7 |$
2) $| 10 | - | -2 |$
3) $| -12 + 1 \times (-5) |$
4) $| 7 - 5 | + | -9 \times (-3) |$
5) $| (14 \div (-7)) \times 3 | - | -2|$
6) $| -36 \div 2^2 + 3 \times (-8) |$
7) $| -6 \times (-3) - 4 \times (-3) + 3^2 |$
8) $\frac{|-15 \div 5 \times 6 + 2|}{|-4^2|}$
9) $| - [(-8 \div 4 + 3 \times 5) - (2^3 + 4)] + 5|$
10) $\frac{|-6| \times |12 - 20| \div |-(-3)|}{-|9 - (-3)|}$

Ordering Integers

Order each set of numbers from least to greatest.

1) $7, -99, 21, -2, 143, -5$
2) $0, -(-9), 120, -1000, |-4|, -|4|$
3) $|2 - 3|, 52, -8 + 5, -1, -3^2, -(-(-70))$
4) $-33, -6 \times 6, -(2 - 2), |5| - |9|, -8^0, 1$
5) $|6 - 9 + 3|, -(12 - 3), -|-|5||, 3 - 3^3, 250, -|16| + |-7|$

Prime and Composite Numbers

Check if the following numbers are prime or composite.

1) 84
2) 71
3) 201
4) 159
5) 173
6) 67
7) 198
8) 207
9) 3,060
10) 311

Factoring Numbers

Factor each number completely into its prime factors.

1) 60
2) 100
3) −144
4) 191
5) −210
6) 105
7) −425
8) 184
9) −905
10) 1,024

Greatest Common Factor (GCF)

Find the GCF.

1) 12 and 18
2) 9 and 10
3) 27 and −63
4) 44 and 55
5) −27 and 81
6) 54, 72 and 18
7) −25, −100 and 120
8) 180, 90 and 75
9) 15, 16, −17 and 18
10) 32, −28, 14 and −20

Least Common Multiple (LCM)

Find the LCM.

1) 8 and 9
2) 10 and 15
3) 20 and 21
4) 2, 3 and 5
5) −25 and 40
6) 45 and 75
7) 11, −10 and 12
8) −15, −30 and −60
9) 18, 24 and 40
10) 36, −15, 24 and 48

Word Problems

Solve the problems.

1) A football team gains 7 yards on the first play, loses 5 yards on the second play, and gains 3 yards on the third play. What is the team's net yardage?

2) A diver descends 10 feet below sea level, then rises 4 feet, and descends another 7 feet. What is the diver's final position relative to sea level?

3) The temperature was $-5°C$ in the morning. It increased by $12°C$ during the day and then dropped by $7°C$ at night. What was the temperature at night?

4) Jacob bought 4 items that each cost $5, and then returned one item for a refund. How much did Jacob spend in total?

5) Maria has 24 red marbles and 36 blue marbles. She wants to create identical sets using all the marbles. What is the greatest number of identical sets she can create, and how many marbles of each color will each set contain?

6) Two friends, Alex and Ben, are running around a track. Alex takes 6 minutes to complete one lap, and Ben takes 9 minutes to complete one lap. After how many minutes will they both be at the starting point together?

7) Jamie has two pieces of ribbon, one 60 inches long and the other 84 inches long. She wants to cut both ribbons into pieces of equal length with no leftover pieces. What is the greatest length of each piece?

8) Let's take 12 noon as our reference point. Hours before noon can be represented with negative integers, while hours after noon can be represented with positive integers. Now, convert the given times into their corresponding integer values

   a. 2 hours before noon
   b. $5:00'\ Am$
   c. $19:00'$
   d. $10:00'\ Pm$

9) Two traffic lights change at different intervals. The first light changes every 6 minutes and the second light change every 9 minutes. If they both change at $8:00\ AM$, at what time will they change next together?

10) The temperature in city A is 8 degrees above zero. The temperature in city B is 10 degrees colder than in city A and 1 degree warmer than in city C. Write the temperatures of all three cities as integers and then calculate the average temperature of each city.

Answer of Worksheets

Integers and opposite Integers (Additive Inverse)
1) −1250
2) −300
3) 250
4) 625
5) Not integer
6) Integer
7) Integer
8) Not integer
9) −11
10) −1

Adding and Subtracting Integers
1) −4
2) 9
3) 0
4) 0
5) −20
6) 41
7) −3
8) −8
9) −3
10) −11

Multiplying and Dividing Integers
1) −30
2) −6
3) −48
4) 10
5) −14
6) 5
7) −2
8) −80
9) −60
10) −1,440

Order of Operations
1) 11
2) 16
3) 6
4) 7
5) 3
6) 43
7) 27
8) 7
9) −20
10) −114

Operation With Absolute Value
1) 4
2) 8
3) 17
4) 29
5) 4
6) 33
7) 39
8) 1
9) 4
10) $\frac{-4}{3}$

Ordering Integers
1) $-99 < -5 < -2 < 7 < 21 < 143$
2) $-1000 < -|4| < 0 < |-4| < -(-9) < 120$
3) $-(-(-70)) < -3^2 < -8 + 5 < -1 < |2-3| < 52$
4) $-6 \times 6 < -33 < |5| - |9| < -8^0 < -(2-2) < 1$
5) $3 - 3^3 < -|16| + |-7| = -(12-3) < -|-|5|| < |6-9+3| < 250$

Prime and Composite Numbers
1) Composite
2) Prime

3) Composite
4) Composite
5) Prime
6) Prime

7) Composite
8) Composite
9) Composite
10) Prime

Factoring Numbers
1) $2^2 \times 3 \times 5$
2) $5^2 \times 2^2$
3) $-1 \times 2^4 \times 3^2$
4) 191
5) $-1 \times 2 \times 3 \times 5 \times 7$
6) $3 \times 5 \times 7$
7) $-1 \times 5^2 \times 17$
8) $2^3 \times 23$
9) $-1 \times 5 \times 181$
10) 2^{10}

Greatest Common Factor (GCF)
1) 6
2) 1
3) 9
4) 11
5) 27
6) 18
7) 5
8) 15
9) 1
10) 2

Least Common Multiple (LCM)
1) 72
2) 30
3) 420
4) 30
5) 200
6) 225
7) 660
8) 60
9) 360
10) 720

Word Problems
1) 5 yards
2) 13 feet below sea level
3) $0°C$
4) $15
5) 12 identical sets, each containing 2 red marbles and 3 blue marbles
6) After 18 minutes
7) 12 inches
8) $a: -2, b: -7, c: +7, d: +10$
9) $8:18\ AM$
10) $1°C$

Chapter 2: Decimals

Topics that you'll learn in this chapter:

- ✓ Comparing Decimals
- ✓ Adding and Subtracting Decimals
- ✓ Multiplying and Dividing Decimals
- ✓ Multiplying Decimals by Power of 10
- ✓ Dividing Decimals by Power of 10
- ✓ Rounding Decimals
- ✓ Word Problems
- ✓ Worksheets
- ✓ Answer of Worksheets

Comparing Decimals

Definition of Decimals:
Decimals are a way of representing numbers that are not whole numbers. They use a decimal point to separate the whole number part from the fractional part. Decimals are an extension of the base-10 number system, which is also known as the decimal system.

Key Points about Decimals:

1. **Decimal Point**:
   - The dot in a decimal number is called the decimal point.
   - It separates the whole number part from the fractional part.

2. **Place Value**:
   - Each digit in a decimal number has a place value.
   - The place value of digits to the left of the decimal point is 1, 10, 100, etc.
   - The place value of digits to the right of the decimal point is 0.1, 0.01, 00.1, etc.

How to order decimals:

1. **Compare digit by digit from left to right:** Start with the leftmost digit (ones place), then move to the right (tenths place, hundredths place, etc.).

2. **Add zeros if necessary:** If the numbers have different lengths, you can add zeros to the end of the shorter decimals to make them the same length. This won't change the value of the number but makes it easier to compare.

Examples:

1) Identify the place values of 45.678.

   Solution:

   4 (tens), 5 (ones), 6 (tenths), 7 (hundredths), 8 (thousandths)

2) Order following decimal numbers from smallest to largest:

$$1.85, 0.197, 1.099, 0.21, 1.645, 0.3$$

   Solution:

   Compare digit by digit from left to right:
   - Start with the smallest from the 0-group: 0.197
   - Follow with the rest from the 0-group: 0.21, 0.3
   - Proceed to the 1-group: 1.099, 1.645, 1.85

   So, the numbers in order from smallest to largest are: $0.197 < 0.21 < 0.3 < 1.099 < 1.645 < 1.85$.

www.mathnotion.com

Adding and Subtracting Decimals

Steps to Add or Subtract Decimals:

1. **Align the Decimal Points**: Write the numbers in a column, aligning the decimal points.

   ☑ For subtracting, make sure the larger number is on top.

2. **Add Zeros if Necessary**: Make sure each number has the same number of digits after the decimal point by adding zeros if needed.

3. **Add/Subtract the Numbers**: Starting from the rightmost digit (the smallest place value), add/subtract the digits in each column.

   ☑ For adding, if a column adds up to more than 9, carry over to the next column on the left.
   ☑ For subtracting, if the digit in the top number is smaller than the digit in the bottom number, borrow from the next column to the left.

4. **Place the Decimal Point**: In the result, place the decimal point directly below the aligned decimal points in the numbers being added.

Examples:

1) Add 1.975 and 10.25.
   Solution:
   Step 1: Align the decimal point:

   10.25
   +1.975
   _____

   Step 2: Add zero to the end of 10.25:

   10.250
   +1.975
   _____

   Step 3: Add numbers normally and place the decimal point:

   10.250
   +1.975
   _____
   12.225

2) Subtract 18.3 and 7.238.
   Solution:
   Step 1: Align the decimal point:

   18.3
   −7.238
   _____

   Step 2: Add zero to the end of 18.3:

   18.300
   −7.238
   _____

   Step 3: Subtract numbers normally and add the decimal point:

   18.300
   −7.238
   _____
   11.062

Multiplying and Dividing Decimals

Steps to Multiply Decimals:

1. **Ignore the Decimals:** Temporarily ignore the decimal points and multiply the numbers as if they were whole numbers.
2. **Multiply:** Perform the multiplication as you normally would with whole numbers.
3. **Count the Decimal Places:** Count the total number of decimal places in both of the original numbers. This means you count the digits to the right of the decimal points.
4. **Place the Decimal Point:** In the product (result), place the decimal point so that it has the same number of decimal places as the total counted in step 3.

Steps to Divide Decimals

1. **Make the Divisor a Whole Number:** If the number you're dividing by (the divisor) is not a whole number, multiply both the divisor and the dividend (the number being divided) by 10, 100, 1000, etc., until the divisor is a whole number. This won't change the value of the division, just how it looks.
2. **Place the Decimal Point in the Quotient:** Directly above the decimal point in the dividend.
3. **Divide as Usual:** Perform the division as you would with whole numbers.
4. **Annex Zeros if Necessary:** If you have remaining digits after dividing and need to continue, add zeros to the right of the dividend and continue the division process.

Examples:

1) Multiply 5.3 and 3.14.
   Solution:
   1. Ignore the Decimals: Treat them as 25 and 314
   2. Multiply: $53 \times 314 = 16{,}642$
   3. Count the Decimal Places: 5.3 has 1 decimal place and 3.14 has 2 decimal places. Total decimal places $= 1 + 2 = 3$
   4. Place the Decimal Point: $5.3 \times 3.14 = 16.642$

$$\begin{array}{r} 314 \\ \times\ 53 \\ \hline 1942 \\ +15700 \\ \hline 16642 \end{array}$$

2) Divide 6.25 and 2.5.
   Solution:
   1. Make the Divisor a Whole Number: Multiply both the divisor (2.5) and the dividend (6.25) by 10. This gives us $62.5 \div 25$.
   2. Place the Decimal Point: We set up the division with the decimal point directly above its original position.
   3. Divide as Usual: $6.25 \div 25 = 2.5$ (as if dividing whole numbers).

$$25\overline{)62.5} \longrightarrow \begin{array}{r} 2. \\ 25\overline{)62.5} \\ -50 \\ \hline 12.5 \end{array} \longrightarrow \begin{array}{r} 2.5 \\ 25\overline{)62.5} \\ -50 \\ \hline 12.5 \\ -12.5 \\ \hline 00.0 \end{array}$$

Multiplying Decimals by Power of Ten

Steps for Multiplying by Positive Powers of Ten:
1. Identify the power of 10 you are multiplying by.
2. Move the decimal point to the right by the number of zeros in the power of 10.
   - ☑ If there are not enough digits to move the decimal point to the required number of places, add zeros to the right.

Steps for Multiplying by Negative Powers of Ten:
1. Identify the power of 10 you are multiplying by.
2. Move the decimal point to the left by the number of places indicated by the power of 10.
   - ☑ If there are not enough digits to move the decimal point to the required number of places, add zeros to the left of the original number.

Example: Multiply.
   a) 9.416×10^2
   b) 5.78×10^{-3}

Solution a: Step 1: Identify the power of 10: $10^2 = 100$

Step 2: Move the decimal point 2 places to the right $9.416 \times 10^2 = 9.416 \times 100 = 941.6$

Solution b: Step 1: Identify the power of 10: $10^{-3} = 0.001$

Step 2: Move the decimal point 3 places to the left: $5.78 \times 10^{-3} = 5.78 \times 0.001 = 0.00578$

Dividing Decimals by Power of Ten

Steps for Dividing by Positive Powers of Ten:
1. Identify the power of 10 you are dividing by.
2. Move the decimal point to the left by the number of zeros in the power of 10.
   - ☑ If there are not enough digits to move the decimal point to the required number of places, add zeros to the left of the original number.

Steps for Dividing by negative Powers of Ten:
1. Identify the power of 10 you are dividing by.
2. Move the decimal point to the right by the number of places indicated by the power of 10.
   - ☑ If there are not enough digits to move the decimal point to the required number of places, add zeros to the right.

Example: Divide.
   a) $103.72 \div 10^3$
   b) $0.638 \div 10^{-2}$

Solution a: Step 1: Identify the power of 10: $10^3 = 1000$

Step 2: Move the decimal point 2 places to the left: $103.72 \div 10^3 = 103.72 \div 1000 = 0.10372$

Solution b: Step 1: Identify the power of 10: $10^{-2} = 0.01$

Step 2: Move the decimal point 2 places to the left: $0.638 \div 10^{-2} = 0.638 \div 0.01 = 63.8$

Rounding Decimals

Rounding decimals can be quite handy when you want to simplify a number to make it easier to work with or understand. Here's how to round decimals, step-by-step:

Steps to Round Decimals:

1. **Identify the place value you want to round to**: This could be the nearest whole number, tenths place, hundredths place, etc.

2. **Look at the digit to the right of the place value you're rounding to**:
   - If this digit is 5 or greater, round up the digit in the place you're rounding to.
   - If this digit is less than 5, keep the digit in the place you're rounding to the same.

3. **Change all the digits to the right of the place you're rounding to, to zero (if dealing with whole numbers) or simply drop them (if dealing with decimals)**.

Examples:

1) Round 16.297 to the nearest tenth.
   Solution:
   Step 1: Identify the place value: Round to the nearest tenth (first digit after the decimal point is 2).
   Step 2: look at the next digit: Tenths place is 2, next digit is 9
   Step 3: Apply rounding rules: Since the next digit (9) is greater than 5, round up:
   16.297 → 16.3

2) Round 321.49 to the nearest whole number.
   Solution:
   Step 1: Identify the place value: Round to the nearest whole number (the digit before the decimal point is 1).
   Step 2: Look at the next digit: Next digit after 1 is 4
   Step 3: Apply rounding rules: Since the next digit (4) is less than 5, round down:
   321.49 → 321

3) Round 0.999 to the nearest hundredth:
   Solution:
   Step 1: Identify the place value: The hundredth place is the second digit to the right of the decimal point. So, for 0.999, the hundredths place is the second 9.
   Step 2: Look at the next digit: This is the thousandths place which is also 9
   Step 3: Apply rounding rules: Since the next digit (9) is greater than 5, round up. Rounding up 9 results in 10, which means we need to carry 1 to the tenths place:
   0.999 → 1.00

Word Problems

Solving word problems involving decimals can be simplified by following a clear, step-by-step approach:

Steps to Solve Word Problems Involving Decimals

1. **Read the Problem Carefully**: Understand what the problem is asking. Look for keywords and identify the relevant information.

2. **Identify the Numbers and Operations**: Determine the decimals involved and the operations needed (addition, subtraction, multiplication, division).

3. **Write Down What You Know**: Make a note of the information given and what you need to find out. This helps organize your thoughts.

4. **Set Up the Equation**: Use the information from the word problem to create a mathematical equation.

5. **Solve the Equation**: Perform the necessary arithmetic operations with decimals. Remember to align decimal points when adding or subtracting and adjust the decimal point correctly when multiplying or dividing.

6. **Check Your Work**: Verify your calculations and make sure your answer makes sense in the context of the problem.

Example:

Emma is hosting a party and needs to buy ingredients for a cake. She needs 2.75 kg of flour, 1.5 kg of sugar, and 0.85 kg of cocoa powder. Flour costs $1.20 per kg, sugar costs $1.50 per kg, and cocoa powder costs $3.00 per kg. How much will Emma spend in total on these ingredients?

Solution:

1. Read the problem: Emma needs to buy flour, sugar and cocoa powder. We need to find the total cost.

2. Identify the numbers and operations: Flour: 2.75 kg at $1.20 per kg, Sugar: 1.5 kg at $1.50 per kg and Cocoa powder: 0.85 kg at $3.00 per kg

    Operation: Multiplication for each item, then addition of the totals

3. Write down what you know: Flour cost: 2.75 kg× $1.20/kg, Sugar cost: 1.5 kg× $1.50/kg and Cocoa powder cost: 0.85 kg× $3.00/kg

4. Set up the equations and solve:
    - Flour: $2.75 \times 1.20 = 3.30$
    - Sugar: $1.5 \times 1.50 = 2.25$
    - Cocoa powder: $0.85 \times 3.00 = 2.55$

5. Add the totals: $3.30 + $2.25 + $2.55 = $8.1

Worksheets

⚑Comparing Decimals

Order following decimals from least to greatest:
1) 0.29, 1.2, 1.1999, 0.3, 0.405
2) 5.001, 50.01, 5.1, 5, 50.1
3) 10.01, 10.002, 10.4, 10.00, 10.39
4) 69, −68.299, 68.78, 67.59, −68.3
5) π, 3.14, 3.15, 3.2, 3.1339

⚑Adding and Subtracting Decimals

Add and subtract.
1) $1.68 - 0.99$
2) $5 + 7.23$
3) $31.25 - 2.74 + 11$
4) $2.74 + 1.552 - 3.66$
5) $9 - 7.85$
6) $10.1 - 8.741 + 3.21$
7) $4.1 - (1.23 + 0.667)$
8) $-(6 - 3.142) + (-3.65 + 8)$
9) $-8.13 + 15 - 3.26$
10) $-(-(8.19)) - 6.32 - 0.056$

⚑Multiplying and Dividing Decimals

Multiply and divide.
1) 0.36×1.2
2) 0.08×0.2
3) 9.1×0.11
4) $0.6 \times 0.02 \times 5$
5) $0.8 \times 1.6 \times 100$
6) $7.2 \div 8$
7) $0.15 \div 0.05$
8) $1.6 \div 7$
9) $28.9 \div 9.3$
10) $0.17 \div 2.3$

⚑Multiplying Decimals by Power of 10

Multiply.
1) 5.81×10
2) $0.247 \times 1,000$
3) 100×0.3
4) 93.7×10^3
5) 0.00015×10^2
6) 0.16×0.01
7) 14.85×0.1
8) $10^{-4} \times 124.2$
9) $10 \times 0.58 \times 10^{-2}$
10) $0.001 \times 100 \times 10^{-1} \times 10,000$

⚑Dividing Decimals by Power of 10

Divide.
1) $89.26 \div 10$
2) $0.56 \div 100$
3) $14 \div 0.01$
4) $0.785 \div 0.001$
5) $754.2 \div 0.1$
6) $715 \div 10^2$
7) $0.03 \div 10^{-1}$
8) $6,000 \div 100 \div 10^{-3}$
9) $0.0387 \div 10^{-2} \div 10^3$
10) $1,000 \div 10^4 \div 10^{-1}$

www.mathnotion.com

Rounding Decimals

Round each number to the specified place value indicated by the underline:

1) 15.<u>3</u>98
2) <u>1</u>.897
3) 0.00<u>0</u>9
4) 5478.<u>6</u>
5) 2.9<u>9</u>8
6) 1<u>5</u>4.65
7) 2.87<u>1</u>9
8) 31.<u>4</u>5
9) 0.<u>5</u>47
10) <u>9</u>81.21

Word Problems

Solve the problems.

1) A car travels at a speed of 55.8 miles per hour. How far will it travel in 3.5 hours? Round your answer to the nearest tenth.
2) A 5.5 kg bag of rice is divided equally among 4 people. How much rice does each person get? Round your answer to the nearest hundredth.
3) Emily bought 2.4 meters of fabric for $6.25 per meter. How much did she spend on the fabric? Then, she bought an additional 1.5 meters. How much did she spend in total?
4) What is the average of 1.25, 8.19, 0.14, 5, 9.22?
5) A rectangular plot of land measures 25.5 meters by 18.75 meters. Calculate the area. If 400 square meters are used for a garden, how much area remains for other uses?
6) A recipe requires 0.75 cups of sugar for each batch of cookies. If you have 6.5 cups of sugar, how many batches of cookies can you make? If each batch makes 12 cookies, how many cookies can you make in total?
7) A tank is filled with 35.75 liters of water. If 12.5 liters are used, and then 8.75 liters are added, and finally the tank is divided into 5 equal parts, how much water does each part contain?
8) If the area of a rectangle is 20.1 square meters and its width is 3 meters, what is its perimeter?
9) If the average of 5 numbers is 7.45, and 3 of them are 5.14, 8.74, and 7.33, what is the sum of the other two numbers?
10) A farmer wants to distribute 375.75 kg of wheat evenly into 15 sacks. He then wants to sell each sack of wheat for $22.50 per kg.
    - How much wheat does each sack contain?
    - How much revenue will he generate from selling all the sacks of wheat?

Answer of Worksheets

Comparing Decimals
1) $0.29 < 0.3 < 0.405 < 1.1999 < 1.2$
2) $5 < 5.001 < 5.1 < 50.01 < 50.1$
3) $10.00 < 10.002 < 10.01 < 10.39 < 10.4$
4) $-68.3 < -68.299 < 67.59 < 68.78 < 69$
5) $3.1339 < 3.14 < \pi < 3.15 < 3.2$

Adding and Subtracting Decimals
1) 0.69
2) 12.23
3) 39.51
4) 0.632
5) 1.15
6) 4.569
7) 2.203
8) 1.492
9) 3.61
10) 1.814

Multiplying and Dividing Decimals
1) 0.432
2) 0.016
3) 1.001
4) 0.06
5) 128
6) 0.9
7) 3
8) $Q: 0.2$ and $R: 0.2$
9) $Q: 3.1$ and $R: 0.07$
10) $Q: 0.07$ and $R: 0.009$

Multiplying Decimals by Power of 10
1) 58.1
2) 247
3) 30
4) 93,700
5) 0.015
6) 0.0016
7) 1.458
8) 0.01242
9) 0.058
10) 100

Dividing Decimals by Power of 10
1) 8.926
2) 0.0056
3) 1,400
4) 758
5) 7,542
6) 7.15
7) 0.3
8) 60,000
9) 0.00387
10) 1

Rounding Decimals
1) 15.4
2) 2
3) 0.001
4) 5478.6
5) 3
6) 150
7) 2.872
8) 31.5
9) 0.5
10) 1,000

Word Problems
1) 195.3 miles
2) 1.38 kg
3) $24.38
4) 4.76
5) 78.125 square meters
6) 8 full batches of cookies and 96 cookies in total
7) 6.4 liters
8) 19.4 meters
9) 16.04
10) 25.05 kg of wheat and $8,454.38 in total revenue

www.mathnotion.com

Chapter 3: Fractions

Topics that you'll learn in this chapter:

- ✓ Simplifying Fractions
- ✓ Improper Fractions
- ✓ Least Common Denominator
- ✓ Converting Fractions or Mixed Numbers to Decimals
- ✓ Converting Decimals to Fractions or Mixed Numbers
- ✓ Ordering Mixed Numbers and Improper Fractions
- ✓ Adding and Subtracting (Fractions and Mixed Numbers)
- ✓ Multiplying and Dividing (Fractions and Mixed Numbers)
- ✓ Mixed Operations (Improper, Mixed numbers, and fractions)
- ✓ Complex Fraction Operations
- ✓ Word Problems
- ✓ Worksheets
- ✓ Answer of Worksheets

Simplifying Fractions

Steps to Simplify Fractions:

1. **Find the Greatest Common Divisor (GCD)**: The GCD of two numbers is the largest number that can divide both of them without leaving a remainder.

2. **Divide the Numerator and Denominator by the GCD**.

3. **Check Your Work**: Ensure that the numerator and denominator have no common factors other than 1.

☑ In general, to simplify fractions, we can also use divisibility rules because when the numerator and denominator are large numbers, writing out all the factors might be time consuming.

Examples:

1) Simplify $\frac{32}{24}$.

   Solution:
   1. Find the GCD of 32 and 24:
      - Factors of 32: 1, 2, 4, 8, 16, 32
      - Factors of 24: 1, 2, 3, 4, 6, 8, 12, 24
      - The GCD is 8.
   2. Divide both the numerator and denominator by GCD:
      $$\frac{32 \div 8}{24 \div 8} = \frac{4}{3}$$
   3. Check your work: 3 and 4 don't have common factors other than 1.

2) Simplify $\frac{105}{165}$.

   Solution:
   1. First, we divide both numbers by 5 (because the unit digit of both numbers is 5):
      $$\frac{105 \div 5}{165 \div 5} = \frac{21}{33}$$
   2. Now we divide 21 and 33 by 3 (because the sum of the digits of both numbers is divisible by 3):
      $$\frac{21 \div 3}{33 \div 3} = \frac{7}{11}$$
   3. Check your work: 7 and 11 don't have common factors other than 1.

Improper Fractions

An improper fraction is a type of fraction where the numerator (the top number) is greater than or equal to the denominator (the bottom number). This means the value of the fraction is equal to or greater than 1, such as $\frac{7}{5}, \frac{15}{2}, \frac{68}{50}, \frac{9}{9}, etc.$

Converting Improper Fractions to Mixed Numbers:

Improper fractions can also be expressed as mixed numbers, which combine a whole number and a proper fraction:

1. **Divide the Numerator by the Denominator**: This will give you a whole number (the quotient) and a remainder.
2. **Write the Quotient as the Whole Number Part**:
3. **Fraction with the Remainder**:
   - The remainder becomes the numerator of the fraction
   - The denominator remains the same as the original fraction
4. **Combine the Whole Number and Fraction**: Put the whole number and the fraction together to form the mixed number.

Converting Mixed numbers to Improper Fractions:

1. **Multiply the Whole Number by the Denominator**: This gives you the total number of parts represented by the whole number.
2. **Add the Numerator to the Result**: Add the result from step 1 to the numerator.
3. **Use the Same Denominator**: The denominator remains the same as in the mixed number.

Examples:

1) Convert $\frac{29}{5}$ to a mixed number.
   Solution:
   1. Divide 29 by 5: The result is 5 with a remainder of 4:
   2. The mixed number will be $5\frac{4}{5}$.

$$\begin{array}{r} 5 \\ 5{\overline{\smash{)}29}} \\ -25 \\ \hline 4 \end{array}$$

2) Convert $4\frac{7}{8}$ to an improper fraction.
   Solution:
   1. Multiply the whole number (4) by the denominator (8): $4 \times 8 = 32$
   2. Add the numerator to the 32: $32 + 7 = 39$
   3. The improper fraction will be: $\frac{39}{8}$

www.mathnotion.com

Least Common Denominator

The Least Common Denominator (LCD) is the smallest multiple that two or more numbers (denominators) have in common. It's particularly useful in solving problems involving fractions and common denominators.

Steps to Find the LCD:

1. **List the Multiples**: Write down a few multiples of each denominator.

2. **Identify the Common Multiples**: Look for multiples that appear in all lists.

3. **Choose the Smallest Common Multiple**: The LCD is the smallest number that appears in all lists.

☑ The prime factorization method is particularly useful when dealing with large denominators, as it saves time compared to listing out all their multiples:
   - Identify all the prime factors that appear in any of the factorizations.
   - For each prime factor, take the highest power that appears in any of the factorizations.
   - Multiply these highest powers together to get the LCD.

Examples:

1) Find the LCD of $\frac{1}{9}, \frac{5}{8}$ and $\frac{11}{12}$.

   Solution:
   1. List the multiples of denominators:
      - Multiples of 8: 8, 16, 24, 32, 40, 48, 56, 64, 72, 80 ...
      - Multiples of 9: 9, 18, 27, 36, 45, 54, 63, 72, 81 ...
      - Multiples of 12: 12, 24, 36, 48, 60, 72 ...
   2. Identify the common multiples: 72, 144 ...
   3. Choose the smallest common multiples: The smallest common multiple is 72. So, the LCD will be 72.

2) Find the LCD of $\frac{211}{360}$ and $\frac{85}{84}$.

   Solution:
   Since 360 and 84 are large denominators, its better to use prime factorization to find LCD of their:
   - The prime factor of 360: $360 = 2^3 \times 3^2 \times 5$
   - The prime factor of 84: $84 = 2^2 \times 3 \times 7$

   So, LCD will be: $2^3 \times 3^2 \times 5 \times 7 = 2,520$.

www.mathnotion.com

Converting to Decimals

Converting Fractions to Decimals: Divide the Numerator by the Denominator.

Converting Mixed Numbers to Decimals:

1. Separate the Whole Number and the Fraction.
2. Convert the Fraction to a Decimal: Divide the numerator by the denominator.
3. Add the Whole Number and the Decimal: Combine the whole number and the decimal value.

Example:

Convert $\frac{7}{8}$ and $3\frac{5}{6}$ to decimals.

Solution:

1. Fraction: $\frac{7}{8} = 7 \div 8 = 0.875$
2. Mixed number: Separate the whole number and fraction: 3 and $\frac{5}{6}$
   - Convert the fraction: $\frac{5}{6} = 5 \div 6 \approx 0.83$
   - Add the whole number and decimal: $3\frac{5}{6} = 3 + 0.83 = 3.83$.

Converting to Fractions or Mixed Numbers

Converting Decimals to Fractions:

1. Write down the decimal as a fraction with 1 as the denominator.
2. Multiply both the numerator and denominator by 10 for each digit after the decimal point to eliminate the decimal.
3. Simplify the fraction by dividing both the numerator and denominator by their greatest common divisor (GCD)

Converting Decimals to Mixed Numbers:

1. Separate the whole number part from the decimal part.
2. Convert the decimal part to a fraction.
3. Combine the whole number and the fraction.

Example: Convert 0.13 to fraction and 2.109 to mixed numbers.

Solution:

1. According to steps above for converting decimal to fraction: $0.13 = \frac{0.13}{1} = \frac{0.13 \times 100}{1 \times 100} = \frac{13}{100}$
2. According to the steps above for converting decimal to mixed numbers: $2.109 = 2 + 0.109$
   $0.109 = \frac{0.109}{1} = \frac{0.109 \times 1000}{1 \times 1000} = \frac{109}{1000}$ so, $2.109 = 2 + \frac{109}{1000} = 2\frac{109}{1000}$

Ordering Mixed Numbers and Improper Fractions

Ordering mixed numbers and improper fractions involves converting them to a common format, usually either all to improper fractions or all to decimals, to make comparisons easier.

Methods to Order Mixed Numbers and Improper Fractions:

1. **Find a Common Denominator** (optional but helpful for fractions): Find the least common multiple (LCM) of the denominators and then compare numerators and order them.
2. **Convert Improper Fractions to Decimals** (optional but helpful for ordering): Divide the numerator by the denominator and then compare decimal numbers.
   ☑ Sometimes converting an improper fraction to a mixed number makes the process easier.

Examples:

1) Order following fractions from least to greatest:
$\frac{9}{5}, \frac{21}{6}, 1\frac{7}{10}, \frac{41}{20}, \frac{16}{30}$

Solution:

1. First, we convert improper fractions to mixed numbers to compare fractions by their whole part:
   - $\frac{9}{5} = 1\frac{4}{5}$
   - $\frac{21}{6} = 3\frac{3}{6}$
   - $\frac{41}{20} = 2\frac{1}{20}$
2. $1\frac{7}{10}$ and $1\frac{4}{5}$ have equal whole parts, so for comparing them we use finding common denominator method: The LCM between 5 and 10 is 10, so $1\frac{4}{5} = 1\frac{8}{10}$
3. Compare all fractions: By looking at their whole part we have:
   $\frac{16}{30} < 1\frac{7}{10} < \frac{9}{5} = 1\frac{8}{10} < \frac{41}{20} = 2\frac{1}{20} < \frac{21}{6} = 3\frac{3}{6}$

2) Order the following fraction from least to greatest:
$\frac{49}{35}, \frac{13}{8}, 1\frac{5}{10}, 1\frac{23}{29}, \frac{11}{9}$

Solution:

If we convert all improper fractions to mixed numbers, we find that the whole part of all of them is one. Therefore, in this case, converting them all to decimals is more logical:

- $\frac{49}{35} = 49 \div 35 = 1.4$
- $\frac{13}{8} = 13 \div 8 = 1.625$
- $1\frac{5}{10} = \frac{15}{10} = 15 \div 10 = 1.5$
- $1\frac{23}{29} = \frac{52}{29} = 52 \div 29 \approx 1.793$
- $\frac{11}{9} = 11 \div 9 \approx 1.22$

According to the results, we have: $\frac{11}{9} < \frac{49}{35} < 1\frac{5}{10} < \frac{13}{8} < 1\frac{23}{29}$.

Adding and Subtracting

Adding and Subtracting Fractions:

1. **Find a Common Denominator:** Find the Least Common Denominator (LCD)
2. **Convert Fractions to Equivalent Fractions:** Convert each fraction to have the common denominator.
3. **Add or Subtract the Numerators:** Add or subtract the numerators while keeping the common denominator.

Adding and Subtracting Mixed Numbers:

1. **Separate the Whole Numbers and Fractions.**
2. **Add or Subtract the Whole Numbers.**
3. **Add or Subtract the Fractions (using the steps above):** Find the common denominator and convert the fractions.
4. **Combine the Whole Number and Fraction.**

Example:

Do the following additions and subtractions:

a) $\frac{5}{9} - \frac{1}{6} + \frac{1}{3}$

b) $5\frac{3}{20} + 1\frac{7}{30}$

Solution a:

1. Find a common denominator for all fractions. The least common denominator (LCD) of 9, 6, and 3 is 18.
2. Convert each fraction to have the common denominator:
   - For $\frac{5}{9}$: $\frac{5}{9} = \frac{5 \times 2}{9 \times 2} = \frac{10}{18}$
   - For $\frac{1}{6}$: $\frac{1}{6} = \frac{1 \times 3}{6 \times 3} = \frac{3}{18}$
   - For $\frac{1}{3}$: $\frac{1}{3} = \frac{1 \times 6}{3 \times 6} = \frac{6}{18}$
3. Perform the addition and subtraction: First subtract $\frac{10}{18}$ and $\frac{3}{18}$, then add the result to $\frac{6}{18}$:

   $\frac{5}{9} - \frac{1}{6} + \frac{1}{3} = \frac{10}{18} - \frac{3}{18} + \frac{6}{18} = \frac{13}{18}$

Solution b:

1. Separate the whole numbers and fractions: Whole numbers are 5 and 1 and fractions are $\frac{3}{20}$ and $\frac{7}{30}$.
2. Add the whole numbers: $5 + 1 = 6$
3. Add the fractions: Find the common denominator and convert the fractions:

   $\frac{3}{20} + \frac{7}{30} = \frac{9}{60} + \frac{14}{60} = \frac{23}{60}$
4. Combine the whole number and fraction: The result is $6\frac{23}{60}$.

www.mathnotion.com

Multiplying and Dividing

Multiplying Fractions and Mixed Numbers:

1. **Convert Mixed Numbers to Improper Fractions**
2. **Multiply the Fractions:** Multiply the numerators and multiply the denominators.
3. **Write the Result as a Fraction:** Combine the products of the numerators and denominators:
4. **Simplify if Needed:** Reducing them to their simplest form and then convert them back to a mixed number if necessary.

Dividing Fractions and Mixed Numbers:

1. **Convert Mixed Numbers to Improper Fractions**
2. **Flip the Second Fraction (Reciprocal)**
3. **Multiply the First Fraction by the Reciprocal of the Second Fraction**
4. **Multiply the Fractions:** Multiply the numerators and multiply the denominators.
5. **Simplify if Needed:** Reducing them to their simplest form and then convert them back to a mixed number if necessary.

Example:

Do the following multiplication and division:

a) $1\frac{2}{10} \times \frac{3}{5}$

b) $1\frac{3}{7} \div 2\frac{1}{3}$

Solution a:

1. Convert mixed numbers to improper fraction: $1\frac{2}{10} = \frac{12}{10}$
2. Multiply the fractions: $\frac{12}{10} \times \frac{3}{5} = \frac{36}{50}$
3. Simplify: $\frac{36}{50} = \frac{36 \div 2}{50 \div 2} = \frac{18}{25}$

Solution b:

1. Convert mixed numbers to improper fractions:
   - $1\frac{3}{7} = \frac{10}{7}$
   - $2\frac{1}{3} = \frac{7}{3}$
2. Flip the second fraction: $\frac{7}{3} \to \frac{3}{7}$
3. Multiply the $\frac{10}{7}$ by the reciprocal of the $\frac{7}{3}$:
   $1\frac{3}{7} \div 2\frac{1}{3} = \frac{10}{7} \div \frac{7}{3} = \frac{10}{7} \times \frac{3}{7} = \frac{30}{49}$
4. Simplify: $\frac{30}{49}$ is in simplest form.

Mixed Operations

Performing mixed operations (addition, subtraction, multiplication, and division) with fractions, including proper, improper, and mixed numbers, involves a few key steps. Here's a comprehensive guide:

Steps for Mixed Operations:

1. **Follow the Order of Operation:** Use PEMDAS (**P**arentheses/Brackets, **E**xponents, **M**ultiplications and **D**ivisions from left to right, **A**ddition and **S**ubtraction from left to right)
2. **Convert Mixed Numbers to Improper Fractions** (if necessary)
3. **Multiply and Divide Fractions**
4. **Find a Common Denominator (for addition and subtraction)**
   - **Multiplication:** Multiply the numerators and denominators.
   - **Division:** Multiply first fraction by the reciprocal of the second fraction
5. **Simplify the Result** (if necessary)

Example:

Do the following expression:

$$1\frac{2}{9} + \frac{1}{3} \times 2\frac{5}{6} - \left(\frac{1}{4} \div 2 + \frac{1}{2}\right)$$

Solution:

1. Do parenthesis: First do the division in parenthesis, then do the addition (the common denominator between 8 and 2 is 8):
   $$\frac{1}{4} \div 2 + \frac{1}{2} = \frac{1}{4} \times \frac{1}{2} + \frac{1}{2} = \frac{1}{8} + \frac{1}{2} = \frac{1}{8} + \frac{4}{8} = \frac{5}{8}$$
2. Convert $2\frac{5}{6}$ to an improper fraction: $2\frac{5}{6} = \frac{17}{6}$
3. Do the multiplication:
   $$\frac{1}{3} \times 2\frac{5}{6} = \frac{1}{3} \times \frac{17}{6} = \frac{17}{18}$$
4. Add $1\frac{2}{9}$ to the result of multiplication (the common denominator between 9 and 18 is 18):
   $$1\frac{2}{9} + \frac{1}{3} \times 2\frac{5}{6} = 1\frac{2}{9} + \frac{17}{18} = 1\frac{4}{18} + \frac{17}{18} = 1\frac{21}{18} = 2\frac{3}{18}$$
5. Subtract the result within the parentheses and the result of the addition from the left side of the expression:
   $$2\frac{3}{18} - \frac{5}{8} = 2\frac{12}{72} - \frac{45}{72} = \frac{156}{72} - \frac{45}{72} = \frac{111}{72}$$ (the common denominator between 8 and 18 is 72 and for subtracting $2\frac{12}{72}$ and $\frac{45}{72}$ it is better to convert $2\frac{12}{72}$ to an improper fraction because we cannot subtract their numerators).

So, the final result is: $1\frac{2}{9} + \frac{1}{3} \times 2\frac{5}{6} - \left(\frac{1}{4} \div 2 + \frac{1}{2}\right) = \frac{111}{72}$

Complex Fraction Operations

Complex fractions might seem a bit daunting at first but breaking them down into simpler steps makes them manageable. A complex fraction is a fraction where the numerator, denominator, or both contain fractions themselves.

Steps to Simplify Complex Fractions:

1. **Simplify the Numerator and Denominator Separately:** Treat each part of the complex fraction as a separate fraction problem.
2. **Convert Mixed Numbers to Improper Fractions in Both Numerator and Denominator** (if necessary)
3. **Follow the Order of Operation for Numerator and Denominator Separately:** If there are a combination of operations follow PEMDAS.
4. **Divide the Result of Numerator and Denominator.**
5. **Simplify the Resulting Fraction.**

Example:

Simplify $\dfrac{\frac{2}{3}+\frac{1}{4}}{1\frac{1}{2}\times\frac{2}{5}}$

Solution:

1. For the numerator: The common denominator for 3 and 4 is 12:

   $$\frac{2}{3}+\frac{1}{4}=\frac{8}{12}+\frac{3}{12}=\frac{11}{12}$$

2. For the denominator: First convert $1\frac{1}{2}$ to an improper fraction the multiply the result by $\frac{2}{5}$:

   $$1\frac{1}{2}\times\frac{2}{5}=\frac{3}{2}\times\frac{2}{5}=\frac{6}{10}$$

3. Divide the Final Result of Numerator and Denominator:

   $$\dfrac{\frac{2}{3}+\frac{1}{4}}{1\frac{1}{2}\times\frac{2}{5}}=\dfrac{\frac{11}{12}}{\frac{6}{10}}=\frac{11}{12}\div\frac{6}{10}=\frac{11}{12}\times\frac{10}{6}=\frac{110}{72}$$

4. Simplify the Resulting Fraction:

   $$\frac{110}{72}=\frac{110\div 2}{72\div 2}=\frac{55}{36}$$

www.mathnotion.com

Word Problems

Word problems involving fractions can be tackled methodically by following these steps:

Steps to Solve Word Problems About Fractions:

1. **Read the Problem Carefully**: Understand what the problem is asking. Identify the key information and the questions to be answered.

2. **Identify the Fractions and Key Information**: Pick out the fractions given in the problem and any other necessary details.

3. **Decide on the Operation**: Determine whether you need to add, subtract, multiply, or divide the fractions.

4. **Find a Common Denominator** (for addition and subtraction): Convert the fractions to have the same denominator.

5. **Perform the Operation**.

6. **Write the Answer in Context**: Answer the question in the context of the problem.

☑ **Draw a Picture**: Visualizing the fractions can help you understand the problem better.

Example:

The farmer planted wheat on $\frac{1}{3}$ of his land and rice on $\frac{3}{4}$ of the remaining land. What fraction of his land was left unplanted?

Solution:

1. Read the problem carefully and identify the fraction and key information: The farmer has a certain amount of land. He planted:
   - Wheat on $\frac{1}{3}$ of his land.
   - Rice on $\frac{3}{4}$ of the remaining land.

   We need to find out what fraction of his land was left unplanted.

2. Decide on operations and perform them:
   - Calculate the remaining land after planting wheat: $1 - \frac{1}{3} = \frac{3}{3} - \frac{1}{3} = \frac{2}{3}$
   - Calculate the land used for planting rice: $\frac{3}{4}$ of $\frac{2}{3} = \frac{3}{4} \times \frac{2}{3} = \frac{6}{12} = \frac{1}{2}$
   - Calculate the land left unplanted: First we calculate the total amount of land planted by wheat and rice. Then we subtract planted land from the entire land:
     - Planted land: $\frac{1}{3} + \frac{1}{2} = \frac{2}{6} + \frac{3}{6} = \frac{5}{6}$
     - Unplanted land: $1 - \frac{5}{6} = \frac{1}{6}$

3. Write the answer in context: $\frac{1}{6}$ of farmer's land was unplanted.

Worksheets

✑ Simplifying Fractions

Simplify following fractions to its simplest form:

1) $\frac{25}{30}$
2) $\frac{49}{21}$
3) $\frac{81}{120}$
4) $\frac{99}{88}$
5) $\frac{42}{24}$
6) $\frac{69}{36}$
7) $\frac{102}{210}$
8) $\frac{580}{290}$
9) $\frac{198}{306}$
10) $\frac{2500}{3000}$

✑ Improper Fractions

Convert improper fractions to mixed numbers and mixed numbers to improper fractions:

1) $\frac{17}{5}$
2) $\frac{39}{21}$
3) $\frac{45}{9}$
4) $\frac{85}{16}$
5) $\frac{118}{10}$
6) $2\frac{3}{7}$
7) $1\frac{4}{5}$
8) $6\frac{1}{11}$
9) $2\frac{2}{2}$
10) $4\frac{20}{21}$

✑ Least Common Denominator

Find the least common denominator (LCD).

1) $\frac{1}{6}$ and $\frac{2}{3}$
2) $\frac{2}{5}$ and $\frac{1}{4}$
3) $\frac{7}{8}$ and $\frac{5}{6}$
4) $\frac{12}{20}$ and $\frac{1}{15}$
5) $\frac{5}{9}$ and $\frac{7}{12}$
6) $\frac{5}{16}$ and $\frac{14}{24}$
7) $\frac{7}{30}, \frac{1}{10}$ and $\frac{9}{15}$
8) $\frac{1}{45}, \frac{11}{60}$ and $\frac{8}{20}$
9) $\frac{25}{48}, \frac{21}{72}, \frac{15}{24}$ and $\frac{32}{90}$
10) $\frac{16}{210}, \frac{29}{140}, \frac{6}{49}$ and $\frac{14}{20}$

✑ Converting to Decimals

Convert to decimals.

1) $\frac{3}{5}$
2) $\frac{7}{1000}$
3) $\frac{1}{16}$
4) $\frac{9}{11}$
5) $\frac{12}{20}$
6) $1\frac{3}{50}$
7) $11\frac{11}{100}$
8) $4\frac{3}{4}$
9) $10\frac{1}{1000}$
10) $7\frac{5}{13}$

Converting to Fractions or Mixed Numbers

Convert to fractions or mixed numbers.

1) 0.04
2) 0.185
3) 0.006
4) 0.103
5) 0.700
6) 12.12
7) 1.546
8) 21.8
9) 2.22
10) 16.9050

Ordering Mixed Numbers and Improper Fractions

Order following numbers from least to greatest:

1) $1\frac{5}{8}, 1\frac{5}{6}, 1\frac{5}{10}$
2) $\frac{5}{12}, \frac{1}{2}, \frac{9}{6}$
3) $\frac{41}{8}, \frac{7}{5}, \frac{18}{6}, \frac{25}{4}$
4) $1\frac{1}{3}, \frac{14}{6}, 2\frac{3}{10}, \frac{18}{15}$
5) $2\frac{9}{5}, 3, 3\frac{7}{20}, \frac{35}{10}$
6) $5\frac{14}{24}, 5\frac{7}{30}, 5\frac{9}{18}, 5\frac{8}{12}, 5\frac{6}{15}$
7) $\frac{42}{84}, \frac{44}{90}, \frac{20}{38}$
8) $\frac{17}{16}, \frac{21}{20}, \frac{12}{11}, \frac{37}{36}$
9) $1\frac{23}{100}, \frac{17}{25}, 2\frac{1}{4}, \frac{54}{50}, 2\frac{345}{500}$
10) $4\frac{341}{342}, 4\frac{287}{288}, 4\frac{512}{513}$

Adding and Subtracting

Solve.

1) $\frac{2}{3} + \frac{1}{6} =$
2) $1\frac{3}{5} - \frac{1}{10} =$
3) $2\frac{5}{6} - \frac{7}{6} =$
4) $1\frac{3}{7} + 2\frac{5}{8} =$
5) $6 - 1\frac{4}{15} =$
6) $2\frac{1}{12} - 1\frac{5}{6} + \frac{1}{4} =$
7) $5\frac{7}{16} - 4\frac{2}{24} =$
8) $4\frac{1}{8} - 2\frac{5}{9} - 1\frac{1}{12} =$
9) $7 - \frac{5}{9} + 1\frac{2}{10} + 6\frac{15}{10} =$
10) $3\frac{3}{15} + 1\frac{4}{24} - 1\frac{7}{70} =$

Multiplying and Dividing

Solve and simplify the result.

1) $\frac{2}{7} \times \frac{5}{9} =$
2) $1\frac{2}{7} \times \frac{14}{9} =$
3) $\frac{5}{16} \times 2\frac{1}{5} =$
4) $3\frac{6}{8} \times 1\frac{1}{7} =$
5) $1\frac{3}{7} \times 1\frac{2}{5} \times \frac{20}{25} =$
6) $\frac{4}{5} \div \frac{1}{8} =$
7) $1\frac{11}{9} \div \frac{20}{9} =$
8) $\frac{3}{8} \div 2\frac{5}{7} =$
9) $3\frac{7}{9} \div 1\frac{1}{3} =$
10) $\frac{6}{10} \div 1\frac{1}{5} \div \frac{6}{5} =$

Mixed Operations

Find answers.

1) $\frac{1}{9} + \left(\frac{2}{3} - \frac{1}{5}\right) =$
2) $1\frac{1}{4} - \frac{5}{6} \times \frac{12}{10} \div 2 =$

3) $10 - \frac{3}{8} \times 1\frac{1}{5} \div 2 =$

4) $\left(2\frac{3}{4} \div \frac{3}{8}\right) \times \left(\frac{5}{6} \div 1\frac{1}{1}\right) =$

5) $\frac{9}{20} \div \left(6 - 2\frac{1}{7}\right) =$

6) $\left(4\frac{7}{12} - 1\frac{9}{10}\right) \div 2 \times \frac{1}{3} =$

7) $4\frac{1}{4} - 2\frac{1}{3} \times \frac{3}{7} + 1\frac{3}{6} =$

8) $\frac{3}{4} \times \left(\left(1 + \frac{2}{7}\right) - 1\frac{1}{6} \div \frac{7}{6}\right) \div \frac{1}{2} =$

9) $\left(\frac{3}{7} \div \frac{1}{14}\right) \div \left(2 \div 1\frac{2}{3}\right) =$

10) $\left(\left(\frac{5}{100} \times 1\frac{1}{99}\right) + 2 \div 3\right) \div 4 =$

Complex Fraction Operations

Solve.

1) $\dfrac{\frac{1}{3} - \frac{1}{6}}{\frac{2}{5} \div \frac{1}{5}} =$

2) $\dfrac{2\frac{3}{7} \div \frac{1}{14}}{3\frac{1}{5} \times \frac{10}{8}} =$

3) $\dfrac{2}{5\frac{1}{6} - \frac{1}{5} + \frac{7}{15}} =$

4) $\dfrac{\frac{3}{15}}{\frac{9}{20}} \div \dfrac{1 - \frac{7}{8}}{\frac{3}{4}} =$

5) $2 + \dfrac{1}{3 - \dfrac{1}{\frac{1}{3} \div \frac{1}{6}}} =$

Word Problems

Complete the following word problems. You can draw a shape to solve them.

1) Emma has $\frac{3}{4}$ of a pizza and shares $\frac{1}{2}$ of it with her friend. How much pizza does she have left?

2) In a classroom, $\frac{2}{5}$ of the students are boys. If there are 25 students, how many girls are there?

3) If the perimeter of a rectangular garden is $12\frac{3}{4}$ meters and its width is $2\frac{1}{3}$ meters, what is the area of this rectangle in square meters?

4) John planted flowers in $\frac{3}{5}$ of his garden and vegetables in $\frac{1}{4}$ of the remaining part. What fraction of his garden is used for vegetables?

5) During a field trip, $\frac{3}{8}$ of the students visited the museum and $\frac{1}{2}$ of the remaining students visited the zoo. What fraction of the students visited the zoo?

6) The length of a walking path is $24\frac{3}{4}$ kilometers. If Sophie walked $\frac{1}{4}$ of the path yesterday and $\frac{2}{3}$ of the path today, how many kilometers did she walk in total over these two days?

7) A cookie recipe uses $\frac{3}{4}$ cup of flour. If you want to make half the recipe, how much flour do you need?

8) A jar is $\frac{3}{4}$ full of water. If $\frac{2}{5}$ of the water is poured out, what fraction of the jar remains full?

9) A car that had $\frac{1}{6}$ of its tank full went to a gas station and with 14 liters, half of its tank was filled. What is the capacity of the car's tank in liters?

10) If $\frac{1}{3}$ of a tanker of water is poured into 4 containers, and $\frac{1}{4}$ of each container is filled, with each container having a capacity of 20 liters, what is the capacity of the tanker?

Answer of Worksheets

Simplifying Fractions
1) $\frac{5}{6}$
2) $\frac{7}{3}$
3) $\frac{27}{40}$
4) $\frac{9}{8}$
5) $\frac{7}{4}$
6) $\frac{23}{12}$
7) $\frac{17}{35}$
8) 2
9) $\frac{11}{17}$
10) $\frac{5}{6}$

Improper Fractions
1) $3\frac{2}{5}$
2) $1\frac{6}{7}$
3) 5
4) $5\frac{5}{16}$
5) $11\frac{4}{5}$
6) $\frac{17}{7}$
7) $\frac{9}{5}$
8) $\frac{67}{11}$
9) 3
10) $\frac{104}{21}$

Least Common Denominator
1) 6
2) 20
3) 24
4) 60
5) 36
6) 48
7) 30
8) 180
9) 720
10) 2,940

Converting to Decimals
1) 0.6
2) 0.007
3) 0.0625
4) ≈ 0.81
5) 0.6
6) 1.06
7) 11.11
8) 4.75
9) 10.001
10) ≈ 7.38

Converting to Fractions or Mixed Numbers
1) $\frac{4}{100} = \frac{1}{25}$
2) $\frac{185}{1000} = \frac{37}{200}$
3) $\frac{6}{1000} = \frac{3}{500}$
4) $\frac{103}{1000}$
5) $\frac{7}{10}$
6) $12\frac{12}{100} = 12\frac{3}{25}$
7) $1\frac{546}{1000} = 1\frac{273}{500}$
8) $21\frac{8}{10} = 21\frac{4}{5}$
9) $2\frac{22}{100} = 2\frac{11}{50}$
10) $16\frac{905}{1000} = 16\frac{181}{200}$

Ordering Mixed Numbers and Improper Fractions
1) $1\frac{5}{10} < 1\frac{5}{8} < 1\frac{5}{6}$
2) $\frac{5}{12} < \frac{1}{2} < \frac{9}{6}$
3) $\frac{7}{5} < \frac{18}{6} < \frac{41}{8} < \frac{25}{4}$
4) $\frac{18}{15} < 1\frac{1}{3} < 2\frac{3}{10} < \frac{14}{6}$
5) $3 < 3\frac{7}{10} < \frac{35}{10} < 2\frac{9}{5}$
6) $5\frac{7}{30} < 5\frac{6}{15} < 5\frac{9}{18} < 5\frac{14}{24} < 5\frac{8}{12}$
7) $\frac{44}{90} < \frac{42}{84} < \frac{20}{38}$
8) $\frac{37}{36} < \frac{21}{20} < \frac{17}{16} < \frac{12}{11}$

9) $\frac{17}{25} < \frac{54}{50} < 1\frac{23}{100} < 2\frac{1}{4} < 2\frac{345}{500}$

10) $4\frac{287}{288} < 4\frac{341}{342} < 4\frac{512}{513}$

Adding and Subtracting
1) $\frac{5}{6}$
2) $1\frac{1}{2}$
3) $1\frac{2}{3}$
4) $4\frac{3}{56}$
5) $4\frac{11}{15}$
6) $\frac{1}{2}$
7) $1\frac{17}{48}$
8) $\frac{35}{72}$
9) $15\frac{13}{90}$
10) $3\frac{4}{15}$

Multiplying and Dividing
1) $\frac{10}{63}$
2) 2
3) $\frac{11}{16}$
4) $\frac{30}{7}$
5) $\frac{8}{5}$
6) $\frac{32}{5}$
7) 1
8) $\frac{21}{152}$
9) $\frac{17}{6}$
10) $\frac{5}{12}$

Mixed Operations
1) $\frac{26}{45}$
2) $\frac{3}{4}$
3) $9\frac{31}{40}$
4) $3\frac{1}{18}$
5) $\frac{7}{60}$
6) $\frac{161}{360}$
7) $4\frac{3}{4}$
8) $\frac{3}{7}$
9) 5
10) $\frac{67}{396}$

Complex Fraction Operations
1) $\frac{1}{12}$
2) $8\frac{1}{2}$
3) $\frac{60}{163}$
4) $\frac{8}{3}$
5) $\frac{12}{5}$

Word Problems
1) $\frac{3}{8}$ of that pizza.
2) 15 girls.
3) $\frac{679}{72}$ square meters.
4) $\frac{1}{10}$ of his garden.
5) $\frac{5}{16}$ of the students.
6) $22\frac{11}{16}$ kilometers.
7) $\frac{3}{4}$ cup of flour.
8) $\frac{9}{20}$ of jar.
9) 42 liters.
10) 60 litters.

Chapter 4: Rational Numbers

Topics that you'll learn in this chapter:

- ✓ Rational Numbers
- ✓ Ordering Rational Numbers
- ✓ Adding and Subtracting Rational Numbers
- ✓ Multiplying and Dividing Rational Numbers
- ✓ Mixed Operations on Rational Numbers
- ✓ Absolute Value Operation
- ✓ Word Problems
- ✓ Worksheets
- ✓ Answer of Worksheets

Rational Numbers

Rational numbers are numbers that can be expressed as the quotient or fraction $\frac{p}{q}$ of two integers, where p (the numerator) and q (the denominator) are integers, and $q \neq 0$.

Some key characteristics of rational numbers include:
- **Whole numbers** (e.g.., 1, 2, 3) are rational because they can be written as $\frac{1}{1}, \frac{2}{1}, and \frac{3}{1}$, respectively.
- **Fractions** (e.g.., $\frac{3}{4}, \frac{-5}{6}$) are rational numbers because they are ratios of integers.
- **Decimals** that either terminate (e.g., 0.5 or 1.25) or repeat (e.g., $0.333...$ or $0.\overline{3}$) are rational because they can be converted into a fraction form.

In summary, any number that can be expressed as a ratio of two integers is a rational number.

Converting a repeating decimal to rational number:

1. **Let the repeating decimal be represented as x.**
2. **Set up an equation with x representing the decimal.**
3. **Multiply both sides of the equation by a power of 10 to move the decimal point to the right:** The number of times you multiply by 10 depends on how many digits are repeating. If one-digit repeats, multiply by 10; if two digits repeat, multiply by 100, and so on.
4. **Subtract the original equation from this new equation.**
5. **Solve for x.**

Example:

Convert $0.\overline{12}$ to a fraction format of rational number.

Solution:

1. Let $x = 0.\overline{12}$.
2. Multiply both sides of the equation by 100 (because two digits repeat):
$100x = 12.\overline{12}$
3. Subtract the original equation from this new equation:
$100x - x = 12.\overline{12} - 0.\overline{12}$
Simplifying: $99x = 12$
4. Solve for x and simplify that: $x = \frac{12 \div 3}{99 \div 3} = \frac{4}{33}$

Thus, the repeating decimal $0.\overline{12}$ is equal to $\frac{4}{33}$.

www.mathnotion.com

Ordering Rational Numbers

To order rational numbers (arrange them in ascending or descending order), follow these steps:

1. **Convert each rational number to a common form (either fraction or decimal):**
   - If the rational numbers are given as fractions or mixed numbers, you can compare them directly, but it may help to express them as decimals to make comparisons easier.
   - If the rational numbers are given as decimals, you can compare them directly by looking at the digits.
2. **Use a common denominator to compare fractions or mixed numbers:** If you prefer to keep the numbers in fraction form, find the least common denominator (LCD) of the fractions. Convert each fraction to have the LCD as the denominator and then compare the numerators.
3. **In case of negative numbers, remember that the more negative a number is, the smaller it is.**

Example:

Order following rational numbers from least to greatest:

$2.4, \frac{4}{5}, -1\frac{3}{4}, -0.98, \frac{7}{8}, -\frac{9}{4}$

Solution:

We can convert all of them into decimals and then arrange them in order:

- $\frac{4}{5} = 4 \div 5 = 0.8$
- $-1\frac{3}{4} = -\frac{7}{4} = -7 \div 4 = -1.75$
- $\frac{7}{8} = 7 \div 8 = 0.875$
- $-\frac{9}{4} = -9 \div 4 = -2.25$

The positive rational numbers are: $2.4, 0.8$ and 0.875 order them from least to greatest: $0.8 < 0.875 < 2.4$.

The negative rational numbers are: $-0.98, -1.75$ and -2.25 order them from least to greatest: $-2.25 < -1.75 < -0.98$. (Note that on the negative side, the farther we move from zero, the smaller the numbers will become).

Combine both groups of numbers:

$$-2.25 < -1.75 < -0.98 < 0.8 < 0.875 < 2.4$$

$$\text{Or } -\frac{9}{4} < -1\frac{3}{4} < -0.98 < \frac{4}{5} < \frac{7}{8} < 2.4$$

Adding and Subtracting Rational numbers

Adding and subtracting rational numbers involves some basic steps, and it depends on whether the numbers are in fraction form or decimal form.

Adding and Subtracting Rational Numbers in Fraction or Mixed Number Format:

1. Convert Mixed Numbers to Improper Fractions.
2. Find a Common Denominator: Find the Least Common Denominator (LCD)
3. Convert Fractions to Equivalent Fractions: Convert each fraction to have the common denominator.
4. Add or Subtract the Numerators: Add or subtract the numerators while keeping the common denominator.

Adding and Subtracting Rational Numbers in Decimal Format:

1. Align the Decimal Points of the Numbers.
2. Add or Subtract as You Would with Integer Whole Numbers.
3. Place the Decimal Point in Result.

Examples:

1) Evaluate this: $-1\frac{5}{6} + \frac{1}{9}$

   Solution:
   1. Convert mixed numbers to improper fraction:
      $$-1\frac{5}{6} = -\frac{11}{6}$$
   2. Find common denominator: The LCD of $-\frac{11}{6}$ and $\frac{1}{9}$ is 18.
   3. Convert to equivalent fractions:
      $$-\frac{11 \times 3}{6 \times 3} = -\frac{33}{18} \text{ and } \frac{1 \times 2}{9 \times 2} = \frac{2}{18}$$
   4. Add the numbers:
      $$-1\frac{5}{6} + \frac{1}{9} = -\frac{11}{6} + \frac{1}{9} = -\frac{33}{18} + \frac{2}{18} = \frac{-33+2}{18} = \frac{-31}{18} = -1\frac{13}{18}$$

2) Evaluate this: $1.854 - 2.6$

   Solution:
   Because the absolute value of -2.6 is greater than that of 1.854, we first compare the absolute values of the numbers. Then, we subtract 1.854 from 2.6 and finally apply a negative sign to the result:
   So, the final answer will be:
   $1.854 - 2.6 = -0.746$

   $$\begin{array}{r} 2.600 \\ -1.854 \\ \hline 0.746 \end{array}$$

Multiplying and Dividing Rational Numbers

Multiplying Rational Numbers in Fraction and Mixed numbers Format:

1. Convert Mixed Numbers to Improper Fractions.
2. Multiply the Numerators and Denominators.

Multiplying Rational Numbers in Decimal Format: To multiply decimal numbers, you multiply them as usual and count the total number of decimal places in both numbers to place the decimal point in the product.

☑ The sign of the product depends on the signs of the factors being multiplied, just like the rules for multiplying integer numbers.

Dividing Rational Numbers in Fraction and Mixed Numbers Format:

1. Convert Mixed Numbers to Improper Fractions.
2. Multiply the First Fraction by the reciprocal (inverse) of the second fraction.

Dividing Rational Numbers in Decimal Format: To divide decimal numbers, move the decimal point of the divisor (the number you're dividing by) to the right until it becomes a whole number. Then, move the decimal point of the dividend (the number being divided) the same number of places to the right, and perform the division as you would with whole numbers.

☑ The sign of the quotient depends on the signs of the dividend and the divisor, just like the rules for dividing integer numbers.

Examples:

1) Multiply -0.52 and 1.6.
   Solution:
   1. Ignore the signs and perform the multiplication as you would with whole numbers: $52 \times 16 = 832$
   2. Count the number of decimal places in the factors and place the decimal point in the product: 0.52 has 2 decimal places and 1.6 has 1 decimal places, since the total number of decimal places is 3, put the decimal point three places from the right in: 0.832
   3. Determine the sign of the product: The product of a negative number and positive number is negative: $-0.52 \times 1.6 = -0.832$

2) Divide $-1\frac{2}{3}$ by $-\frac{4}{9}$.
   Solution:
   1. Convert mixed numbers to improper fraction: $-1\frac{2}{3} = -\frac{5}{3}$
   2. Multiply the first fraction by the reciprocal of second fraction: $-\frac{5}{3} \div \left(-\frac{4}{9}\right) = -\frac{5}{3} \times \left(-\frac{9}{4}\right)$
   3. Determine the sign of the quotient and simplify the result: When you divide two negative numbers the result is positive: $-\frac{5}{3} \div \left(-\frac{4}{9}\right) = -\frac{5}{3} \times \left(-\frac{9}{4}\right) = \frac{45 \div 3}{12 \div 3} = \frac{15}{4}$

Mixed Operation on Rational Numbers

Mixed operations on rational numbers refer to performing a combination of addition, subtraction, multiplication, and division within the same mathematical expression:

Steps to Perform Mixed Operation:

1) **Convert all Mixed Numbers to Improper Fractions.**
2) **Follow the Order of Operations (PEMDAS):** Parentheses, Exponents, Multiplication and Division (from left to right), Addition and Subtraction (from left to right).
3) **Simplifying at Each Step:** Simplifying makes calculation easier.
- ☑ When both decimals and fractions are present in an expression, consider the following two scenarios:
  - If the denominators of the fractions can easily be converted to a power of 10, you can convert the fractions to decimals and solve the expression.
  - If the denominators cannot be converted to a power of 10, it is better to convert the decimal numbers to fractions and solve the expression.

Examples:

1) Evaluate: $-1\frac{4}{8} \div (\frac{4}{5} \times 2 - \frac{1}{3})$.
   Solution:
   1. Convert mixed numbers to improper fraction: $-1\frac{4}{8} = -\frac{12 \div 4}{8 \div 4} = -\frac{3}{2}$
   2. Parentheses: First, evaluate the expression within the parentheses: First, we do the multiplication and then do the subtraction:
   $\frac{4}{5} \times 2 - \frac{1}{3} = \frac{8}{5} - \frac{1}{3} = \frac{8 \times 3}{5 \times 3} - \frac{1 \times 5}{3 \times 5} = \frac{24}{15} - \frac{5}{15} = \frac{19}{15}$
   3. Do the division: $-\frac{3}{2} \div \frac{19}{15} = -\frac{3}{2} \times \frac{15}{19} = -\frac{45}{38}$

   So, in summary: $-1\frac{4}{8} \div (\frac{4}{5} \times 2 - \frac{1}{3}) = -\frac{3}{2} \div (\frac{8}{5} - \frac{1}{3}) = -\frac{3}{2} \div \frac{19}{15} = -\frac{3}{2} \times \frac{15}{19} = -\frac{45}{38}$

2) Evaluate $\frac{1}{15} - 0.12 \times \frac{1}{2} + 1.25$
   Solution:
   1. Convert decimals to fractions: Because the denominator of $\frac{7}{15}$ is not converted to power of ten so, it's better to convert decimals to fractions:
   $0.12 = \frac{12 \div 4}{100 \div 4} = \frac{3}{25}$ and $1.25 = \frac{125 \div 25}{100 \div 25} = \frac{5}{4}$
   2. Do the multiplication: $0.12 \times \frac{1}{2} = \frac{3}{25} \times \frac{1}{2} = \frac{3}{50}$
   3. Do the subtraction: The common denominator is 150: $\frac{1}{15} - \frac{3}{50} = \frac{10}{150} - \frac{9}{150} = \frac{1}{150}$
   4. Do the addition: The common denominator is 300: $\frac{1}{150} + \frac{5}{4} = \frac{2}{300} + \frac{375}{300} = \frac{377}{300}$

   So, in summary: $\frac{1}{15} - 0.12 \times \frac{1}{2} + 1.25 = \frac{1}{15} - \frac{3}{25} \times \frac{1}{2} + \frac{5}{4} = \frac{1}{15} - \frac{3}{50} + \frac{5}{4} = \frac{1}{150} + \frac{5}{4} = \frac{377}{300}$

Absolute Value Operation

The absolute value of a rational number is its distance from zero on the number line, regardless of its sign. In other words, the absolute value of a number is always non-negative.

For a rational number a, the absolute value is denoted as $|a|$, and it is defined as:

$$|a| = \begin{cases} a & \text{if } a \geq 0 \\ -a & \text{if } a < 0 \end{cases}$$

Operation Involving Absolute Values:

- **Addition and Subtraction:** First, we must perform the operations inside the absolute value and then determine the absolute value of the result, and we should note that:
  $|a + b| \neq |a| + |b|$ and $|a - b| \neq |a| - |b|$
- **Multiplication:** The absolute value of the product of two numbers is the product of their absolute values:
  $|a \times b| = |a| \times |b|$
- **Division:** The absolute value of the quotient is the quotient of the absolute values:
  $|a \div b| = |a| \div |b|$

Examples:

1) Find the absolute value of the following rational numbers:
   $-5\frac{2}{3}, 0.16, -\left(-\frac{7}{18}\right), -1.542$

   Solution:
   - $\left|-5\frac{2}{3}\right| = 5\frac{2}{3}$
   - $|0.16| = 0.16$
   - $\left|-\left(-\frac{7}{18}\right)\right| = \frac{7}{18}$
   - $|-1.542| = 1.542$

2) Do the following expressions:
   a) $\left|-1\frac{2}{5} + \frac{3}{10}\right|$
   b) $\left|\frac{6}{20} - (-0.8)\right|$
   c) $\left|-\frac{14}{30} \times \left(-\frac{9}{7}\right)\right|$
   d) $|5 \div -(-2.6)|$

   Solution:

   a) $\left|-1\frac{2}{5} + \frac{3}{10}\right| = \left|-\frac{7}{5} + \frac{3}{10}\right| = \left|-\frac{14}{10} + \frac{3}{10}\right| = \left|-\frac{11}{10}\right| = \frac{11}{10}$

   b) $\left|\frac{6}{20} - (-0.8)\right| = \left|\frac{6}{20} + 0.8\right| = \left|\frac{3}{10} + 0.8\right| = |0.3 + 0.8| = |1.1| = 1.1$

   c) $\left|-\frac{14}{30} \times \left(-\frac{9}{7}\right)\right| = \left|-\frac{14}{30}\right| \times \left|-\frac{9}{7}\right| = \frac{14 \div 7}{30 \div 3} \times \frac{9 \div 3}{7 \div 7} = \frac{2 \div 2}{10 \div 2} \times \frac{3}{1} = \frac{1}{5} \times \frac{3}{1} = \frac{3}{5}$

   d) $|5 \div -(-2.6)| = |5| \div |-(-2.6)| = 5 \div 2.6 = \frac{5}{1} \div \frac{26}{10} = \frac{5}{1} \times \frac{10}{26} = \frac{50 \div 2}{26 \div 2} = \frac{25}{13}$

Word Problems

Tackling word problems involving rational numbers can be very rewarding once you get the hang of it! Here are some steps to help you solve them effectively:

Steps to Solve Word Problems with Rational Numbers

1. **Read the Problem Carefully**: Understand what the problem is asking for. Identify key information and any rational numbers involved.

2. **Identify the Operations**: Determine which mathematical operations (addition, subtraction, multiplication, division) are needed to solve the problem.

3. **Translate Words to Numbers**: Convert the word problem into a mathematical expression or equation using the rational numbers provided.

4. **Solve the Equation**: Perform the necessary calculations using the rational numbers.

5. **Check Your Work**: Verify your solution by plugging it back into the context of the problem to see if it makes sense.

Example:

Amir owns a small bookstore. He tracks the daily sales to monitor the store's performance. On Monday, Amir's bookstore sold books and made a profit of 20.18 dollars. However, on Tuesday, he had to give a refund for a returned book, which amounted to $-20\frac{1}{5}$ dollars. What is Amir's total profit/loss over these two days?

Solution:

1. Read about the problem carefully: Amir had a profit of 20.18 dollars on Monday and a loss amounted to $-20\frac{1}{5}$ dollars on Tuesday.

2. Identify the operation: We must add 20.18 and $-20\frac{1}{5}$ to find if there was a profit or a loss.

3. Translate to numbers: $20.18 + (-20\frac{1}{5})$

4. Solve the equation:

$$20.18 + \left(-20\frac{1}{5}\right) = 20.18 - 20\frac{1\times2}{5\times2} = 20.18 - 20\frac{2}{10} = 20.18 - 20.2 = -0.02$$

5. Check your work: He has lost 0.02 dollars or 2 cents over these two days.

Worksheets

🚩Rational Numbers
Convert to fraction.
1) $0.\overline{3}$
2) $0.\overline{23}$
3) $1.\overline{4}$
4) $21.1\overline{6}$
5) $3.1\overline{25}$
6) $11.00\overline{4}$

🚩Ordering Rational Numbers
Order the rational numbers.
1) $-\frac{2}{5}, -\frac{2}{3}, -\frac{2}{13}$
2) $1.3, -\frac{5}{12}, \frac{1}{5}, -1\frac{3}{5}, -1.2$
3) $-2\frac{1}{7}, 0.89, -0.9, -\frac{11}{25}$
4) $3\frac{5}{25}, -14, 10.52, -\frac{1}{1000}, \frac{18}{5}$
5) $0.2, -1.3, -1\frac{1}{3}, \frac{13}{10}, -\frac{16}{5}$
6) $1.02, -3.14, \frac{20}{11}, -\frac{38}{35}, -2.21$

🚩Adding and Subtracting Rational Numbers
Add or subtract.
1) $-\frac{7}{8} + \frac{1}{3} =$
2) $2.8 - 7.36 =$
3) $-3\frac{4}{5} - 2 =$
4) $\frac{2}{5} + 1\frac{5}{6} - 2.1 =$
5) $-\frac{4}{5} - 3\frac{1}{12} + \frac{7}{30} =$
6) $-7 + 8.251 =$
7) $1\frac{1}{2} - 0.8 + 4 =$
8) $2.15 - 1\frac{4}{10} - 3.4 =$
9) $2.3 - \left(\frac{1}{16} - \frac{5}{6}\right) =$
10) $-\left(-\frac{7}{20} + 2\frac{1}{10} - 3\frac{8}{15}\right) + 6 =$

🚩Multiplying and Dividing Rational Numbers
Do the operations.
1) $\frac{9}{20} \times \left(-\frac{7}{8}\right) =$
2) $-\frac{12}{7} \times 2\frac{1}{3} =$
3) $-3.15 \times (-0.2) =$
4) $-\frac{1}{2} \times 0.3 \times 4 =$
5) $-2\frac{3}{6} \div \frac{1}{2} =$
6) $\frac{3}{7} \div \left(-1\frac{1}{2}\right) =$
7) $-4.8 \div 12 =$
8) $-\frac{3}{4} \div (-0.25) \times 0.4 =$
9) $\frac{3}{8} \times (-0.7) \div \frac{1}{10} =$
10) $1.8 \div \left(-\frac{9}{10} \times (-0.2)\right) =$

🚩Mixed Operation on Rational Numbers
Find answers.
1) $-1\frac{1}{2} - \frac{1}{5} \times 2 =$
2) $3 \div \left(-\frac{2}{3} - \frac{1}{5}\right) =$
3) $-\frac{1}{4} \times 0.2 - 4.5 \times 0.1 =$
4) $-5.6 \div (-1.2 + 0.5) =$
5) $-3.2 + 1.8 \times 0.2 \div 0.5 =$
6) $-2\frac{3}{5} + 1\frac{1}{2} \div \frac{2}{3} \times 0.4 =$
7) $-2.2 \times \left(-1\frac{1}{5} - 1\frac{3}{8}\right) \div \frac{1}{3} =$

www.mathnotion.com

7th Grade Rhode Island Math

8) $\dfrac{-1\frac{1}{6}-\frac{1}{2}}{0.3 \div \frac{1}{5}} =$

10) $-\dfrac{\frac{3}{4}-\left(-\frac{5}{9}\right)}{-1\frac{1}{6} \div 2} - \dfrac{1}{3-\frac{-3}{1-\frac{1}{2}}} =$

9) $5.4 - \dfrac{-2.5}{1-\frac{3}{10} \times \frac{1}{2}} =$

🕮 Absolute Value of Rational Numbers and Operation on Them

Solve.

1) $|-0.25| =$

2) $-\left|-\dfrac{5}{12}\right| =$

3) $-\left|-\left(-\dfrac{1}{15}\right) - \dfrac{2}{3}\right| =$

4) $|-3.65 - (-0.8) - 0.4| =$

5) $\left|-\dfrac{7}{25} \times \dfrac{1}{10} - 2\right| =$

6) $-|-0.3| - |4.15| =$

7) $\left|1\dfrac{3}{20} \div 3\dfrac{2}{5}\right| + \left|-\dfrac{1}{10}\right| =$

8) $\dfrac{|2-(-8)| \div |-(-2)|}{-|2.2 \times 0.5|} =$

9) $\left|-\dfrac{7}{8} \div \dfrac{-2}{4}\right| \times \left(-\left|-\dfrac{1}{3}\right|\right) =$

10) $|-0.105 \div 0.3| - |-1.8 \times 0.2| =$

🕮 Word Problems

Solve the problems.

1) The recipe calls for 1.25 cups of sugar. If you only have $\frac{2}{3}$ cup, how much more sugar do you need?

2) Tom has $10. He saves $\frac{1}{4}$ of it every week. How much money does he save in one month?

3) A car uses 8.4 gallons of gas for every 120 miles. How many gallons of gas will it use for a 180-mile trip?

4) A rope is cut into pieces of $\frac{3}{8}$ meter each. If the total length of the rope is 2.25 meters, how many pieces are there?

5) Amy has $29.15 and went to a stationery store. She picked up 3 pens at $0.45 each, 6 notebooks at $3 each, 5 markers at $1.5 each, and a book for $15. Can she buy all these items? If not, how much is she short?

6) We made a square with a perimeter of 3.78 decimeters using a metal wire. If we want to use the same wire to make an equilateral triangle, what will the length of each side in decimeters?

7) In a game, points scored are shown as positive and points lost are shown as negative. If Sarah scored the following points in order: $3.25, \dfrac{-1}{4}, -1\dfrac{3}{4}, -2.75,$ and 4, what is her total score?

8) Write 4 rational numbers between $-2\frac{1}{3}$ and $-3\frac{1}{4}$.

9) Find the average of Peter's grades in the 5 subjects: Mathematics (85.2), Literature (70.8), Chemistry (65), Physics (92.4) and Biology (75.3).

10) Two cars are 180 kilometers apart and are moving towards each other. If the first car travels $\frac{2}{5}$ of the distance and the second car travels $\frac{1}{4}$ of the distance in one hour, how far apart will they be after two hours?

www.mathnotion.com

Answer of Worksheets

Rational Numbers
1) $\frac{1}{3}$
2) $\frac{23}{99}$
3) $\frac{13}{9}$
4) $\frac{127}{6}$
5) $\frac{1547}{495}$
6) $\frac{2476}{225}$

Ordering Rational Numbers
1) $-\frac{2}{3} < -\frac{2}{5} < -\frac{2}{13}$
2) $-1\frac{3}{5} < -1.2 < -\frac{5}{12} < \frac{1}{5} < 1.3$
3) $-2\frac{1}{7} < -0.9 < -\frac{11}{25} < 0.89$
4) $-14 < -\frac{1}{1000} < 3\frac{5}{25} < \frac{18}{5} < 10.52$
5) $-\frac{16}{5} < -1\frac{1}{3} < -1.3 < 0.2 < \frac{13}{10}$
6) $-3.14 < -2.21 < -\frac{38}{35} < 1.02 < \frac{20}{11}$

Adding and Subtracting Rational Numbers
1) $-\frac{13}{24}$
2) -4.56
3) $-5\frac{4}{5}$
4) $\frac{2}{15}$
5) $-3\frac{13}{20}$
6) 1.251
7) 4.7
8) -2.65
9) $\frac{737}{240}$
10) $7\frac{47}{60}$

Multiplying and Dividing Rational Numbers
1) $\frac{-63}{160}$
2) -4
3) 0.63
4) -0.6
5) -5
6) $-\frac{2}{7}$
7) -0.4
8) 1.2
9) $-\frac{21}{8}$
10) 10

Mixed Operation on Rational Numbers
1) $-\frac{19}{10}$
2) $\frac{-45}{13}$
3) -0.5
4) 8
5) -2.48
6) $-\frac{17}{10}$
7) $\frac{3399}{200}$
8) $\frac{-10}{9}$
9) $\frac{709}{85}$
10) $\frac{-4}{9}$

Absolute Value of Rational Numbers and Operation on Them
1) 0.25
2) $-\frac{5}{12}$
3) $-\frac{3}{5}$
4) 3.25
5) $\frac{507}{250}$
6) -4.45
7) $\frac{149}{340}$
8) $\frac{-50}{11}$
9) $\frac{-7}{192}$
10) -0.01

Word Problems
1) $\frac{7}{12} \approx 0.58$ more cups of sugar
2) $10
3) 12.6 gallons of gas
4) 6 pieces
5) Amy is short $12.70 and cannot buy all the items
6) 1.26 decimeters
7) 2.5 points
8) $-3.2, -3, -2.8$ and -2.5
9) 77.74
10) 54 kilometers

Chapter 5: Proportions, Ratios and Percent

Topics that you'll learn in this chapter:

- ✓ Simplifying Ratios
- ✓ Understanding Proportions
- ✓ Solving Proportions
- ✓ Similarity and Ratios
- ✓ Ratios and Rates Word Problems
- ✓ Percentage Calculations
- ✓ Discount, Tax and Tip
- ✓ Simple Interest
- ✓ Compound Interest
- ✓ Word Problems
- ✓ Worksheets
- ✓ Answer of Worksheets

Simplifying Ratios

A ratio is a way to compare two quantities by showing the relative size of one quantity to the other. It expresses how many times one value contains or is contained within the other. Ratios can be written in several forms:

- **Using a colon:** $3:1$
- **As a fraction:** $\frac{3}{1}$
- **In words:** $3\ to\ 1$

Simplifying Ratios

Simplifying ratios involves reducing the terms of the ratio to their smallest whole numbers while maintaining the same proportional relationship. This process is similar to simplifying fractions.

Steps to Simplify Ratios:

1. **Identify the Terms of the Ratio:** Write the ratio in the form a: b.
2. **Find the greatest common divisor (GCD) of the terms:** The GCD is the largest number that divides both terms without leaving a remainder.
   - ☑ **For Ratios with Variables:** look for common factors in the numerator and the denominator.
3. **Divide both terms by the GCD:** This will give you a simplified ratio.
   - ☑ **For Ratios with Variables:** Divide both the numerator and the denominator by their greatest common factor.

Examples:

1) Simplify the ratio $8:12$.
   Solution:
   1. Identify the terms: 8 and 12.
   2. Find the GCD of 8 and 12: The GCD is 4.
   3. Divide both terms by the GCD: $\frac{8 \div 4}{12 \div 4} = \frac{2}{3} = 2:3$

2) Simplify the ratio $15x^3y^2 : 25xy^3$.
   Solution:
   1. Identify the terms: $15x^3y^2$ and $25xy^3$.
   2. Find the GCD of the coefficients and common factors in numerator and the denominator : The GCD and common factor is $5xy^2$.
   3. Simplify the variables: $\frac{15x^3y^2}{25xy^3} = \frac{3x^2}{5y} = 3x^2:5y$

   Therefore, the simplified ratio is $3x^2:5y$.

Understanding Proportions

Definition: Two ratios are said to be proportional if they represent the same relationship between numbers. In other words, two ratios $\frac{a}{b}$ and $\frac{c}{d}$ are proportional if $\frac{a}{b} = \frac{c}{d}$. This can also be written as $a : b = c : d$.

Identifying Proportional Ratios

To determine if two ratios are proportional, you can use the following methods:

1. **Cross-Multiplication:** If the cross-products of the ratios are equal, then the ratios are proportional. For $\frac{a}{b}$ and $\frac{c}{d}$, check if $a \times d = b \times c$.

2. **Simplifying Ratios:** Simplify both ratios to their lowest terms. If the simplified ratios are the same, then the original ratios are proportional.

Examples:

1) Determine if the ratios $\frac{2}{3}$ and $\frac{4}{6}$ are proportional (use Cross-Multiplication method).
   Solution:
   Cross-Multiplication: $2 \times 6 = 12$ and $3 \times 4 = 12$
   Since $12 = 12$, the ratios $\frac{2}{3}$ and $\frac{4}{6}$ are proportional.

2) Determine if the $\frac{30}{36}$ and $\frac{40}{48}$ are proportional. (use Simplifying Ratio method)
   Solution:
   Simplify both ratios:
   - $\frac{30 \div 6}{36 \div 6} = \frac{5}{6}$
   - $\frac{40 \div 8}{48 \div 8} = \frac{5}{6}$

   Since the simplified forms of these two ratios are not equal, the two ratios are not proportional.

3) Which of following ratios are proportional with the ratio $\frac{12}{9}$?

   a) $\frac{16}{20}$ b) $\frac{20}{15}$ c) $\frac{21}{35}$ d) $\frac{12}{27}$

   Solution:
   First simplify the original ratio: $\frac{12 \div 3}{9 \div 3} = \frac{4}{3}$

   Now simplify the other four ratios:

   a) $\frac{16 \div 4}{20 \div 4} = \frac{4}{5}$ b) $\frac{20 \div 5}{15 \div 5} = \frac{4}{3}$ c) $\frac{21 \div 7}{35 \div 7} = \frac{3}{5}$ d) $\frac{12 \div 3}{27 \div 3} = \frac{4}{9}$

   so, the ratios $\frac{12}{9}$ and $\frac{20}{15}$ are proportional.

Solving Proportions

Solving proportions involves finding the value of a variable that makes two ratios equal. Here's a step-by-step guide to help you solve proportions:

Step-by-Step Guide

1. **Set up the Proportion:** Write the proportion as two equal ratios, where one ratio contains the unknown variable. For example, if we have the proportion:

$$\frac{x}{b} = \frac{c}{d}$$

2. **Cross-Multiply:** Multiply the numerator of one ratio by the denominator of the other ratio. This gives you two cross-products:

$$x \times d = b \times c$$

3. **Solve for the Unknown Variable:** If the proportion contains an unknown variable, isolate the variable on one side of the equation:

$$x = \frac{b \times c}{d}$$

Examples:

1) Solve the proportion: $\frac{12}{9} = \frac{x}{15}$

   Solution:
   - Cross-multiply:
     $12 \times 15 = 9 \times x$
     $180 = 9x$
   - Solve for x: $x = \frac{180}{9} = 20$

2) To make a type of orange color, we mix red and yellow colors in a ratio of 18 to 20. If we use 30 units of yellow color, how many units of red color should we use to make the same orange color?

   Solution:
   1. Identify the ratio: The given ratio of red to yellow color is 18 to 20.
   2. Set up the proportion: We can write the ratio as: $\frac{Red}{Yellow} = \frac{18}{20}$
   3. Set up proportion equation: $\frac{18}{20} = \frac{x}{30}$
   4. Cross-multiply to solve for x: $18 \times 30 = 20 \times x$
   5. Calculate the cross-products: $540 = 20x$
   6. Solve for x: $x = \frac{540}{20} = 27$

   So, to make the same orange color, you need 27 units of red color when using 30 units of yellow color.

Similarity and Ratios

Similarity in geometry refers to the relationship between two shapes that have the same shape but may differ in size. The two shapes are similar if their corresponding angles are equal, and their corresponding sides are proportional.

Key Points of Similarity

1. **Corresponding Angles**: Equal.
2. **Corresponding Sides**: Proportional.

Example of Similarity:

If triangle $\triangle ABC$ is similar to triangle $\triangle DEF$, then:

1. **Angles**: $\angle A = \angle D, \quad \angle B = \angle E, \quad \angle C = \angle F$
2. **Sides**: $\frac{AB}{DE} = \frac{BC}{EF} = \frac{CA}{FD}$

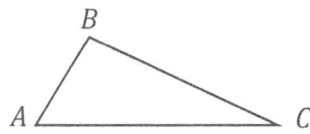
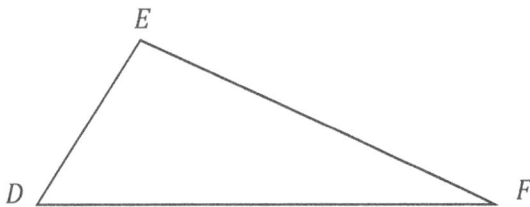

Relationship between Similarity and Ratios:

When two shapes are similar, the ratios of the lengths of their corresponding sides are equal. This means that the concept of ratios is essential in proving similarity between geometric figures.

Example:

Consider two similar triangles, $\triangle ABC$ and $\triangle DEF$. Side lengths of $\triangle ABC$ are: $AB = 6, BC = 8$ and $CA = 10$. In $\triangle DEF$, side $DE = 9$. Find the lengths of EF and FD in $\triangle DEF$.

Solution:

1. Set up the proportional relationship: $\frac{AB}{DE} = \frac{BC}{EF} = \frac{CA}{FD}$
2. Using the given side lengths: $\frac{6}{9} = \frac{8}{EF} = \frac{10}{FD}$
3. Simplify the ratio: $\frac{6}{9} = \frac{2}{3}$
4. Find the length of EF: $\frac{2}{3} = \frac{8}{EF} \rightarrow 2 \times EF = 3 \times 8 \rightarrow 2EF = 24 \rightarrow EF = \frac{24}{2} = 12$
5. Find the length of FD: $\frac{2}{3} = \frac{10}{FD} \rightarrow 2 \times FD = 3 \times 10 \rightarrow 2FD = 30 \rightarrow FD = \frac{30}{2} = 15$

So, the lengths of sides EF and FD in $\triangle DEF$ are 12 and 15 respectively.

Ratios and Rates Word Problems

Differences Between Ratios and Rates

1. **Ratios:**

    - **Definition:** A ratio is a comparison of two quantities that have the same unit. It shows the relative size of one quantity to the other.
    - **Example:** The ratio of 3 apples to 4 oranges can be written as $3:4$ or $\frac{3}{4}$.

2. **Rates:**

    - **Definition:** A rate is a comparison of two quantities that have different units. It shows the relative size of one quantity per unit of the other quantity.
    - **Example:** A car travels 60 miles in 1 hour, which can be written as 60 miles per hour (60 mph).

Solving Word Problems about ratios and rates: Read carefully, set up proportions, solve for the unknown, and check your work.

Examples:

1) The ratio of boys to girls in a class is $5:4$. If there are 15 boys. How many girls are there?
   Solution:
   1. Set up the ratio: $\frac{boys}{girls} = \frac{5}{4}$
   2. Identify the known quantity: There are 15 boys.
   3. Set up the proportion: $\frac{5}{4} = \frac{15}{x}$
   4. Cross-multiply: $5 \times x = 4 \times 15 \rightarrow 5x = 60$
   5. Solve for x: $x = \frac{60}{5} = 12$

   So, there are 12 girls in the class.

2) A factory produces 240 widgets in 8 hours. How many widgets does the factory produce per hour?
   Solution:
   1. Identify the quantities:
      - Total number of widgets = 240
      - Total time in hours = 8
   2. Set up the rate:
      $$Rate\ of\ production = \frac{Total\ number\ of\ widgets}{Total\ time}$$
   3. Calculate the rate:
      $$Rate\ of\ production = \frac{240\ widgets}{8\ hours} = 30\ widgets\ per\ hour$$

   So, the factory produces 30 widgets per hour.

www.mathnotion.com

Percentage Calculations

Percentage calculations involve finding the ratio of a number to 100. Percentages are used to express how one quantity compares to another as a fraction of 100.

Common Percentage Calculations

1. **Finding the Percentage of a Number:** To find $p\%$ of a number N, use the formula:
$Percentage\ of\ N = \frac{p}{100} \times N$

2. **Finding the Percentage a Part of a Whole:** To find the percentage a part A of a whole B use this formula:
$Percentage = (\frac{Part\ (A)}{Whole\ (B)}) \times 100$

3. **Finding the Whole, given a Part and a Percentage:** To find a whole (B), given a part (A) and a percentage use this formula:
$Whole\ (B) = (\frac{Part\ (A)}{percentage}) \times 100$

4. **Increasing or Decreasing a Number by a Percentage:**
Formula for Increase: $New\ Value:\ Original\ Value + (\frac{Percentage}{100} \times Original\ Value)$
Formula for Decrease: $New\ Value:\ Original\ Value - (\frac{Percentage}{100} \times Original\ Value)$

Examples:

1) What is 20% of 50?
Solution:

To find 20% of 50, we use the following formula:

$Percentage\ of\ N = \frac{p}{100} \times N \rightarrow 20\%\ of\ 50 = \frac{20}{100} \times 50 = 0.2 \times 50 = 10$

2) What percentage of 200 is 40?
Solution:
To find the percentage a part 40 of a whole 200 use this formula:

$Percentage = (\frac{Part\ (A)}{Whole\ (B)}) \times 100 = \frac{40}{200} \times 100 = \frac{4000}{200} = 20$

3) If 60 is the 30% of a number, find the number.
Solution:
To find a whole, given a part (60) and a percentage (30%) use this formula:

$Whole\ (B) = (\frac{Part\ (A)}{Percentage}) \times 100 = \frac{60}{30} \times 100 = 2 \times 100 = 200$

4) Increase 80 by 25%.
Solution:
$New\ Value: = 80 + (\frac{25}{100} \times 80) = 80 + (0.25 \times 80) = 80 + 20 = 100$

5) Decrease 48 by 10%.
Solution:
$New\ Value: = 48 - (\frac{10}{100} \times 48) = 48 - (0.1 \times 48) = 48 - 4.8 = 43.2$

Discount, Tax and Tip

Discount, tax, and tip problems are practical applications of percentage calculations. They involve using percentages to calculate the amount to be subtracted (discount), added (tax or tip), and finding the final total.

1. **Discount Problems:** Discount problems involve calculating the amount of money saved on an item when a percentage discount is applied.

   Calculate the Discount and the Sale Price:

   $$Discount = Original\ Price \times \left(\frac{Discount\ Percentage}{100}\right) \text{ and } Sale\ Price = Original\ Price - Discount$$

2. **Tax Problems:** Tax problems involve calculating the amount of tax added to the price of an item based on a percentage tax rate.

   Calculate the Tax and Total Cost:

   $$Tax = Price \times \left(\frac{Tax\ Rate}{100}\right) \text{ and } Total\ Cost = Price + Tax$$

3. **Tip Problems:** Tip problems involve calculating the amount of money added to a bill as a gratuity, based on a percentage of the total bill.

   Calculate the Tip and Total Amount:

   $$Tip = Bill\ Amount \times \left(\frac{Tip\ Percentage}{100}\right) \text{ and } Total\ Amount = Bill\ Amount + Tip$$

Examples:

1) An item costs $50 and is on sale for 20% off. How much is the discount, and what is the sale price?
   Solution:

   $$Discount = 50 \times \left(\frac{20}{100}\right) = 50 \times 0.2 = 10 \quad and \quad Sale\ Price = 50 - 10 = 40$$

   So, the discount is $10 and the sale price is $40.

2) An item costs $100, and the sales tax rate is 8%. How much is the tax, and what is the total cost?
   Solution:

   $$Tax = 100 \times \left(\frac{8}{100}\right) = 100 \times 0.08 = 8 \quad and \quad Total\ Cost = 100 + 8 = 108$$

   So, the tax is $8 and the total cost is $108.

3) The restaurant bill is $75, and you want to leave a 15% tip. How much is the tip, and what is the total amount to be paid?
   Solution:

   $$Tip = 75 \times \left(\frac{15}{100}\right) = 75 \times 0.15 = 11.25 \quad and \quad Total\ Amount = 75 + 11.25 = 86.25$$

   So, the tip is $11.25 and the total amount to be paid is $86.25.

www.mathnotion.com

Simple Interest

Simple interest is a method of calculating the interest charged or earned on a principal amount of money over a period of time. It is called "simple" because the interest is calculated only on the initial principal, not on any interest that has been added to the principal.

Simple Interest Formula: The formula for calculating simple interest is

$$Simple\ Interest\ (SI) = P \times R \times T$$

Where:

- "P" is the principal amount (initial sum of money)
- "R" is the annual interest rate (expressed as a decimal or percentage)
- "T" is the time the money is invested or borrowed for, in years

Examples:

1) Suppose you invest $1,000 at an annual interest rate of 5% for 3 years. How much interest will you earn?

   Solution:

   1. Identify the values:
      - *Principal (P) = $1,000*
      - *Annual interest rate (R) = 5% = 0.05*
      - *Time (T) = 3 years*
   2. Calculate the simple interest:
   $Simple\ Interest\ (SI) = P \times R \times T = \$1{,}000 \times 0.05 \times 3 = \$150$

2) You deposit $2,000 in a saving account that pays an annual simple interest. If you earn $240 interest after 4 years, how much is the annual simple interest rate?

   Solution:

   1. Identify the values:
      - *Principal (P) = $2,000*
      - *Time (T) = 4 years*
      - *Simple interest (SI): $240*
   2. Calculate the annual simple interest rate:
   $Simple\ Interest\ (SI) = P \times R \times T \rightarrow R = \frac{SI}{P \times T} = \frac{240}{2{,}000 \times 4} = \frac{240}{8{,}000} = \frac{3}{100} = 3\%$

   So, the annual simple interest rate is 3%.

Compound Interest

Compound interest is a method of calculating interest where the interest earned or paid is added to the principal amount, so that the interest in the next period is calculated on the new total.

Compound Interest Formula: The formula for compound interest is:

$$A = P(1 + \frac{r}{n})^{nt}$$

Where:

- "A" is the future value of the investment/loan, including interest.
- "P" is the principal investment/loan amount.
- "r" is the annal interest rate (decimal).
- "n" is the number of times that interest is compounded per year.
- "t" is the number of years the money is invested/borrowed for.

Example:

Suppose you invest $1,000 at an annual interest rate of 5%, compounded annually, for 3 years. How much will the investment be worth at the end of 3 years?

Solution:

1. Identity the values:
   - *Principal (P) = $1,000*
   - *Annual interest rate (r) = 5% = 0.05*
   - *Number of times compounded per year (n) = 1 (compounded annually)*
   - *Time (t) = 3 years*

2. Use the compound interest formula:

   $$A = P\left(1 + \frac{r}{n}\right)^{nt} = 1000 \times \left(1 + \frac{0.05}{1}\right)^{1 \times 3} =$$

   $1000 \times (1 + 0.05)^3 = 1000 \times 1.05^3 =$

   $1000 \times 1.157625 = 1157.63$

   So, the investment will be worth $1157.63

Word Problems

Here is a general structure to solve word problems involving proportions, ratios and percentages:

General Structure for Proportions, Ratios and Percentages:

Read the Problem Carefully:

- Understand what is being asked.
- Identify the key quantities involved.
- Note any given ratios, proportions, or percentages.

Set Up the Equation or Ratio:

- Write the two ratios that are being compared and set them equal to each other.
- Write the ratio in the form of a fraction and simplify if possible.
- Use the percentage formula based on what you need to find.

Substitute Known Values: Plug in the values given in the problem into your equation or ratio.

Solve for the Unknown:

- Use algebraic methods to isolate the unknown variable.
- Cross-multiply if solving proportions.
- Perform the necessary arithmetic operations.

Check Your Solution:

- Verify that your answer makes sense in the context of the problem.
- Double-check your calculations to ensure accuracy.

Example:

In school, the ratio of boys to girls is 3: 4. If 20% of the girls participate in a science club and there are 180 boys in total, how many girls in the school participate in the science club?

Solution:

1. Identify the given information:
   - Ratio of boys to girls: 3: 4
   - Total number of boys = 180
   - Percentage of girls participating in the science club= 20%
2. Set up an equation to calculate the total number of girls and solve that:
   $\frac{boys}{girls} = \frac{3}{4} \rightarrow \frac{180}{x} = \frac{3}{4} \rightarrow 180 \times 4 = 3 \times x \rightarrow 720 = 3x \rightarrow x = \frac{720}{3} = 240$
3. Calculate the number of girls participating in the science club:
   $Number\ of\ girls\ in\ the\ science\ club = 20\% \times 240 = \frac{20}{100} \times 240 = 0.20 \times 240 = 48$
   So, 48 girls in the school participate in the science club.

Worksheets

Simplifying Ratios

Simplify.
1) $18:21$
2) $12:30$
3) $45:42$
4) $125:25$
5) $105:201$

Simplify the ratios with variables:
6) $16x^3:12x^2$
7) $20x^2y:4x^3y^2$
8) $24a^4b^2:36b^3$
9) $81z^3x:27z^5y^2x^2$
10) $35x^4y^3:45y^4x^3$

Understanding Proportions

Determine whether the following ratios are proportional or not?
1) $\frac{12}{15}$ and $\frac{16}{20}$
2) $\frac{18}{8}$ and $\frac{19}{9}$
3) $\frac{35}{25}$ and $\frac{70}{50}$
4) $\frac{32}{40}$ and $\frac{24}{35}$
5) $\frac{36}{48}$ and $\frac{15}{18}$

Solving Proportions

Find the value of x.
1) $\frac{7}{20} = \frac{x}{10}$
2) $\frac{32}{22} = \frac{16}{x}$
3) $\frac{x}{16} = \frac{21}{48}$
4) $\frac{10}{x} = \frac{72}{36}$
5) $\frac{18}{45} = \frac{14}{x}$
6) $\frac{x}{30} = \frac{35}{42}$
7) $\frac{1.5}{3.5} = \frac{x}{7}$
8) $\frac{2.6}{x} = \frac{1.3}{3}$
9) $\frac{\frac{2}{8}}{9} = \frac{2}{x}$
10) $\frac{x}{\frac{3}{8}} = \frac{16}{5}$

Similarity and Ratios

The quadrilaterals below are similar, find the requested values.
1) DC
2) BC
3) EF
4) $\angle B$
5) $\angle C$
6) $\angle D$
7) $\angle E$
8) $\angle G$

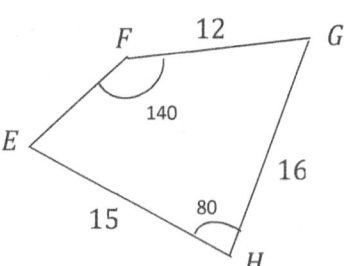

Ratios and Rates Word Problems (Difference)

Solve the problems.

1) A recipe calls for 2 cups of sugar for every 5 cups of flour. If you use 10 cups of flour, how many cups of sugar do you need?
2) A runner completes 3 laps in 15 minutes. At this rate, how long will it take to complete 10 laps?
3) The ratio of cats to dogs in a shelter is 5: 2. If there are 21 cats and dogs in total, how many dogs are there?
4) A recipe requires 3 parts water to 2 parts juice. If you need to make 15 liters of the mixture, how many liters of water and juice do you need?
5) A factory produces 1,200 widgets in 8 hours. How many widgets can it produce in 15 hours at the same rate?

Percentage Calculations

Find the answers.

1) What is 12% of 60?
2) A student scores 60 out of 80 in an exam. What percentage did the student score?
3) A car's value depreciates by 15% each year. If the car's initial value is $20,000, what will its value be after one year?
4) If 40% of a number is 30, what is the whole number?
5) The price of a laptop decreased from $500 to $400. What is the percentage decrease in the price of the laptop?

Discount, Tax and Tip

Solve.

1) If a shirt cost $40 and is on sale for 25% off, what is the sale price?
2) A customer leaves a 20% tip on a $75 restaurant bill. How much is the tip and what is the total bill including the tip?
3) A book is priced at $30, and the sales tax rate is 8%. What is the total cost of the book including tax?
4) If you leave an 18% tip on a $90 restaurant bill, how much do you tip, and what is the total amount paid?
5) A customer buys a dress that is originally priced at $200. The store is offering a 10% discount on the dress. After the discount, a sales tax of 8% is applied to the discounted price. Finally, the customer decides to leave a $5 tip for the service. What is the total amount the customer pays?

Simple Interest

Do the problems.

1) A savings account earns 4% simple interest annually. If you deposit $1,200, how much interest will you have earned after 3 years?

2) If $800 is invested in a savings account with a simple interest rate of 3.5% per year, what will the total amount in the account after 5 years?
3) You want to save $800 in interest over 4 years with an annual interest rate of 3%. How much should you initially invest?
4) A principal amount of $6,500 is invested at a simple interest rate of 3.75% per year. After how many years will the interest earned be equal to $1,950?
5) Alex invested $2,000 in a savings account for 3 years. At the end of the investment period, Alex earned $300 in simple interest. What was the annual simple interest rate?

Compound Interest

Solve (use calculator if necessary).
1) You invest $1,000 in a savings account that pays an annual interest rate of 5%, compounded annually. How much will the investment be worth after 2 years?
2) A principal amount of $2,000 is invested at an annual interest rate of 4%, compounded semi-annually. What will be the total amount after 3 years?
3) An investor deposits $10,000 in an account that pays an annual interest rate of 7%, compounded monthly. How much will the investment be worth after 1 years?
4) Emma wants to have $5,000 in her savings account after 3 years. The account earns an annual interest rate of 4%, compounded annually. What principal amount should Emma deposit now to achieve her goal?
5) David wants his investment of $2,000 to grow to $2,800 in 2 years. The interest is compounded annually. What is the annual interest rate required for David to achieve his goal?

Word Problems

Do the word problems:
1) The ratio of similarity between two rectangles is 3 to 4. If the perimeter of the larger rectangle is 48 centimeters, what is the perimeter of the smaller rectangle in centimeters?
2) Sarah finished a 140-page book in 6 hours, and John finished a 120-page book in 5 hours. Whose reading speed was faster?
3) In a class of 24 students, 18 students received A and B grades in math, and in another class of 30 students, 21 students received A and B grades in the same exam. Which class performed better?
4) If the scale of a map is 1: 40,000 and the length of a part of the map is 3 centimeters, what will be the real length of this part in meters?
5) The population of a town increases by 5% every year. If the current population is 10,000, what will the population be after 3 years?
6) An item priced at $80 is marked down by 25%. After the discount, a sales tax of 7% is applied. What is the final price?
7) A store offers a 30% discount on a $150 item and an additional 10% discount on the already reduced price. What is the final price of the item?

8) A company's revenue increased from $1,000,000 to $1,250,000 in one year. What is the percentage increase in revenue?

9) If we decrease the side of a square by 10%, by what percentage will its area decrease?

10) If a product is taxed once at 5% and then again at 8% on the new price, what is the total percentage tax on the original price?

Answer of Worksheets

Simplifying Ratios
1) $6:7$
2) $2:5$
3) $15:14$
4) $5:1$
5) $35:67$
6) $4x:3$
7) $5:xy$
8) $2a^4:3b$
9) $3:z^2y^2x$
10) $7x:9y$

Understanding Proportions
1) yes
2) no
3) yes
4) no
5) no

Solving Proportions
1) $x = 3.5$
2) $x = 11$
3) $x = 7$
4) $x = 5$
5) $x = 35$
6) $x = 25$
7) $x = 3$
8) $x = 6$
9) $x = 72$
10) $x = 1.2$

Similarity and Ratios
1) 6.4
2) 4.8
3) 5
4) 140°
5) 80°
6) 80°
7) 60°
8) 80°

Ratios and Rates Word Problems (Difference)
1) 4 cups
2) 50 minutes
3) 6 dogs
4) 9 liters of water and 6 liters of juice
5) 2,250 widgets

Percentage Calculations
1) 7.2
2) 75%
3) $17,000
4) 75
5) 20%

Discount, Tax and Tip
1) $30
2) Tip: $15 and Total bill: $90
3) $32.40
4) Tip: $16.50 and Total amount: $106.20
5) $199.40

Simple Interest
1) $144
2) $940
3) $6,666.67
4) ≈ 8 years
5) 5%

Compound Interest
1) $1,102.50
2) ≈ $2,252.32
3) ≈ $10,725
4) ≈ $4,445.57
5) ≈ 18.32%

Word Problems
1) 36 cm
2) John
3) First class
4) 1,200 meters
5) ≈ 11,576
6) $64.20
7) $94.50
8) 25%
9) 19%
10) 13.4%

Chapter 6: Exponents and Radical Expressions

Topics that you'll learn in this chapter:

- ✓ Multiplication Property of Exponents
- ✓ Division Property of Exponents
- ✓ Negative Exponents and Negative Bases
- ✓ Fraction and Decimal Bases
- ✓ Scientific Notation
- ✓ Square Roots
- ✓ Estimate Square Roots
- ✓ Word Problems
- ✓ Worksheets
- ✓ Answer of Worksheets

Multiplication Property of Exponents

1. **Product of Powers Property:**
   Product of powers property states that when you multiply two expressions with the same base, you can add their exponents. Mathematically, this property can be expressed as:
   $$a^n \times a^m = a^{n+m}$$
   Where, a is the base and n, m are the exponents.

2. **Power of a Product Property:**
   When you raise a product to an exponent, you apply the exponent to each factor in the product:
   $$(a \times b)^n = a^n \times b^n$$

3. **Power of Power Property:**
   When you raise an exponent to another exponent, you multiply the exponents:
   $$(a^n)^m = a^{nm}$$

Examples:

1) Simplify $5^3 \times 5^4$.
   Solution:
   Using the product of powers: $5^3 \times 5^4 = 5^7$

2) Simplify $(8 \times 2)^4$.
   Solution:
   Using the power of product property: $(8 \times 2)^4 = 16^4$

3) Simplify $(3^7)^5$.
   Solution:
   Using power of power property: $(3^7)^5 = 3^{35}$

4) Simplify $(x^2 y)^3 \times (x^3 y^2)^2$.
   Solution:
   1. Apply the power of product property inside each term:
      - $(x^2 y)^3 = (x^2)^3 \times y^3$
      - $(x^3 y^2)^2 = (x^3)^2 \times (y^2)^2$

   2. Apply the power of power property inside each term:
      - $(x^2)^3 \times y^3 = x^6 \times y^3$
      - $(x^3)^2 \times (y^2)^2 = x^6 \times y^4$

   3. Combine the results:
      $$(x^2 y)^3 \times (x^3 y^2)^2 = x^6 \times y^3 \times x^6 \times y^4$$

   4. Apply the product of powers property:
      $$(x^2 y)^3 \times (x^3 y^2)^2 = x^6 \times y^3 \times x^6 \times y^4 = x^6 \times x^6 \times y^3 \times y^4 = x^{12} \times y^7$$

Division Property of Exponents

The division properties of exponents help simplify expressions where exponents are involved in division. Here are the key properties:

1. **Quotient of Powers Property:**
   When you divide two expressions with the same base, you subtract the exponents:
   $a^n \div a^m = \frac{a^n}{a^m} = a^{n-m}$

2. **Power of Quotient Property:**
   When you raise a quotient to an exponent, you can apply the exponent to both the numerator and the denominator:
   $(a \div b)^n = \left(\frac{a}{b}\right)^n = \frac{a^n}{b^n}$

3. **Zero Exponent Property:**
   Any non-zero number raised to the power of zero is 1:
   $$a^0 = 1 \ (for \ any \ a \neq 0)$$

Examples:

1) Simplify $\frac{7^6}{7^4}$.

   Solution:

   Using the quotient of power property: $\frac{7^6}{7^4} = 7^{6-4} = 7^2$

2) Simplify $\left(\frac{24}{8}\right)^5$.

   Solution:

   Using the power of a quotient property: $\left(\frac{24}{8}\right)^5 = (24 \div 8)^5 = 3^5$

3) Simplify $\frac{(12^3 \div 3^3)^2 \times 5^{10}}{(6^4 \div 3^4)^2 \times 5^6}$.

   Solution:
   1. Simplify the numerator:
      - Apply the power of a quotient property: $(12^3 \div 3^3)^2 \times 5^{10} = (4^3)^2 \times 5^{10}$
      - Apply the power of power property: $(4^3)^2 \times 5^{10} = 4^6 \times 5^{10}$
      - Write the number 4 as a power with base 2: $4^6 \times 5^{10} = (2^2)^6 \times 5^{10}$
      - Again, apply the power of power property: $(2^2)^6 \times 5^{10} = 2^{12} \times 5^{10}$
   2. Simplify the denominator:
      Apply the power of a quotient property: $(6^4 \div 3^4)^2 \times 5^6 = (2^4)^2 \times 5^6$
      Apply the power of power property: $(2^4)^2 \times 5^6 = 2^8 \times 5^6$
   3. Combine numerator and denominator and apply quotient of power for powers of 2 and 5:
      $$\frac{2^{12} \times 5^{10}}{2^8 \times 5^6} = \frac{2^{12}}{2^8} \times \frac{5^{10}}{5^6} = 2^{12-8} \times 5^{10-6} = 2^4 \times 5^4$$
   4. Apply power of a product property: $2^4 \times 5^4 = 10^4$

www.mathnotion.com

Negative Exponents and Negative Bases

Understanding negative exponents involves knowing how to interpret and compute reciprocals, while negative bases affect the sign of the result based on whether the exponent is even or odd.

Negative Exponents:

A negative exponent indicates that you should take the reciprocal of the base and then apply the positive exponent. In mathematical terms, it's expressed as:

$$a^{-n} = \frac{1}{a^n}$$

Negative Bases:

When dealing with negative bases, the behavior of the exponent depends on whether the exponent is even or odd:

1. **Even Exponent**: When the exponent is even, the result will be positive because multiplying a negative number an even number of times results in a positive product.
2. **Odd Exponent**: When the exponent is odd, the result will be negative because multiplying a negative number an odd number of times results in a negative product.

Combining Negative Exponents and Negative Bases:

You might also encounter scenarios where both the base and the exponent are negative. In such cases, you'll first handle the negative exponent by taking the reciprocal of the base and then apply the exponent.

Examples:

1) Rewrite the negative exponents as positive exponents:

   a) 5^{-2}
   b) $(-2)^{-7}$
   c) $(-7)^{-4}$

   Solution:

   a) $5^{-2} = \frac{1}{5^2}$

   b) $(-2)^{-7} = \frac{1}{-2^7} = -\frac{1}{2^7}$ (the exponent is odd, so the result is negative)

   c) $(-7)^{-4} = \frac{1}{7^4}$ (the exponent is even, so the result is positive)

2) Simplify $\left(\frac{-3a^2}{2b^{-3}}\right)^{-2}$.

   Solution:

   - Simplifying inside the parentheses first:

   $$\left(\frac{-3a^2}{2b^{-3}}\right)^{-2} = \left(\frac{-3a^2 b^3}{2}\right)^{-2}$$

   - Applying the negative exponent outside:

   $$\left(\frac{-3a^2 b^3}{2}\right)^{-2} = \left(\frac{2}{-3a^2 b^3}\right)^2 = \frac{2^2}{(-3a^2 b^3)^2} = \frac{4}{9a^4 b^6}$$

Fraction and Decimal Bases

When dealing with fraction and decimal bases in exponentiation, it's important to understand how to handle the calculations step-by-step.

Fraction Bases:

1. **Raising a Fraction to a Positive Exponent:** $\left(\frac{a}{b}\right)^n = \frac{a^n}{b^n}$
2. **Raising a Fraction to a Negative Exponent:** $\left(\frac{a}{b}\right)^{-n} = \left(\frac{b}{a}\right)^n = \frac{b^n}{a^n}$

Decimal Bases:

1. **Raising a Decimal to a Positive Exponent:** Convert the decimal to a fraction (if necessary), apply the exponent, and then convert back to a decimal.
2. **Raising a Decimal to a Negative Exponent:** Convert the decimal to a fraction, apply the negative exponent, and then convert back to a decimal.

Examples:

1) Simplify the expression $(0.25)^2 \times \left(\frac{3}{4}\right)^{-2}$.

   Solution:
   1. Simplify $(0.25)^2$: $0.25 = \frac{1}{4}$ and $(0.25)^2 = \left(\frac{1}{4}\right)^2 = \frac{1}{4^2}$
   2. Simplify $\left(\frac{3}{4}\right)^{-2} = \left(\frac{4}{3}\right)^2 = \frac{4^2}{3^2}$
   3. Combine the results: $(0.25)^2 \times \left(\frac{3}{4}\right)^{-2} = \frac{1}{4^2} \times \frac{4^2}{3^2} = \frac{1}{3^2} = \frac{1}{9}$

2) Simplify the expression $\frac{\left(-\frac{2}{5}\right)^{-2} \times \left(\frac{1}{3}\right)^{-1}}{3^{-1} \times (0.6 \div 3)^{-2}}$.

   Solution:
   1. Simplify the numerator:
      - Convert negative exponent to positive exponent: $\left(-\frac{2}{5}\right)^{-2} \times \left(\frac{1}{3}\right)^{-1} = \left(\frac{5}{2}\right)^2 \times \left(\frac{3}{1}\right)^1 = \frac{5^2}{2^2} \times 3 = \frac{5^2 \times 3}{2^2}$
   2. Simplify the denominator:
      - Convert negative exponent to positive exponent and apply the power of a quotient property:
      $3^{-1} \times (0.6 \div 3)^{-2} = \frac{1}{3} \times 0.2^{-2} = \frac{1}{3} \times \left(\frac{2}{10}\right)^{-2} = \frac{1}{3^1} \times \left(\frac{10}{2}\right)^2 = \frac{1}{3^5} \times 5^2 = \frac{5^2}{3^1}$
   3. Combine the result of numerator and denominator:
   $\frac{\frac{5^2 \times 3}{2^2}}{\frac{5^2}{3}} = \frac{5^2 \times 3}{2^2} \div \frac{5^2}{3} = \frac{5^2 \times 3}{2^2} \times \frac{3}{5^2} = \frac{5^2 \times 3^2}{2^2 \times 5^2} = \frac{3^2}{2^2} = \left(\frac{3}{2}\right)^2$

   So, the final result is: $\frac{\left(-\frac{2}{5}\right)^{-2} \times \left(\frac{1}{3}\right)^{-1}}{3^{-1} \times (0.6 \div 3)^{-2}} = \left(\frac{3}{2}\right)^2$

Scientific Notation

Scientific notation is a way of writing very large or very small numbers in a compact form. It's especially useful in science and mathematics because it makes calculations easier to handle.

Key Components of Scientific Notation

1. **Coefficient**: A number greater than or equal to 1 and less than 10.
2. **Base**: Always 10.
3. **Exponent**: An integer that shows how many times the base (10) is multiplied by itself.

General Format: The general form of scientific notation is: $a \times 10^n$

Where a is the coefficient. 10 is the base and n is the exponent.

Steps to Convert a Number to Scientific Notation

1. Move the decimal point in the number to the right of the first non-zero digit.
2. Count the number of places you moved to the decimal point. This will be the exponent.
3. If you moved the decimal point to the left, the exponent is positive. If you moved it to the right, the exponent is negative.

Examples:

1) Convert 45,000 to scientific notation.
   Solution:
   1. Move the decimal point: 4.5
   2. Count the places moved: 4 (since 45,000 becomes 4.5)
   3. Write in scientific notation: 4.5×10^4

2) Convert 0.0025 to scientific notation:
   Solution:
   1. Move the decimal point: 2.5
   2. Count the places moved: 3 (since 0.0025 becomes 2.5)
   3. Write in scientific notation: 2.5×10^{-3}

3) Convert $0.0000316 \times 10^7 \times 0.2$ to scientific notation.
   Solution:
   1. Find the product of 0.0000316 and 0.2: $0.0000316 \times 0.2 = 0.00000632$
   2. Move the decimal point: 6.32
   3. Count the places moved: 6 (since 0.00000632 becomes 6.32)
   4. Write in scientific notation: 6.32×10^{-6}
   5. Multiply the result above by 10^7: $6.32 \times 10^{-6} \times 10^7 = 6.32 \times 10^1$

So, the final scientific notation is: $0.0000316 \times 10^7 \times 0.2 = 6.32 \times 10^1$.

www.mathnotion.com

Square Roots

A square root of a number is a value that, when multiplied by itself, gives the original number. It's like asking, "What number can I square (multiply by itself) to get this number?"

Key Concepts:

1. **Symbol**: The square root is represented by the radical symbol: $\sqrt{}$
2. **Positive and Negative Roots**: Every positive number has two square roots, one positive and one negative ($\sqrt{x}$ and $-\sqrt{x}$).
3. **Perfect Squares**: Numbers like $1, 4, 9, 16, 25, etc.$, which are squares of integers $(1, 2, 3, 4, 5, etc.)$, have whole number square roots.
4. **Square Root of Zero**: The square root of zero is zero: $\sqrt{0} = 0$

Square Root Properties and Operations:

1. **Product Property**: $\sqrt{a \times b} = \sqrt{a} \times \sqrt{b}$
2. **Quotient Property**: $\sqrt{\frac{a}{b}} = \frac{\sqrt{a}}{\sqrt{b}}$
3. **Simplifying Square Roots**: To simplify $\sqrt{x}$, factor x into its prime factors and pair the factors.

Example:

Simplify following expressions:

a) $\sqrt{49}$
b) $\sqrt{0.01}$
c) $-\sqrt{36 \times 81}$
d) $\sqrt{\frac{16}{0.25}}$
e) $\frac{\sqrt{72} \times \sqrt{50}}{-\sqrt{0.04}}$

Solution:

a) $\sqrt{49} = 7$ because $7 \times 7 = 49$

b) $\sqrt{0.01} = \sqrt{\frac{1}{100}} = \frac{\sqrt{1}}{\sqrt{100}} = \frac{1}{10} = 0.1$ because $0.1 \times 0.1 = 0.01$

c) $-\sqrt{36 \times 81} = -\sqrt{36} \times \sqrt{81} = -6 \times 9 = -54$

d) $\sqrt{\frac{16}{0.25}} = \frac{\sqrt{16}}{\sqrt{0.25}} = \frac{4}{\sqrt{\frac{25}{100}}} = \frac{4}{\frac{\sqrt{25}}{\sqrt{100}}} = \frac{4}{\frac{5}{10}} = 4 \div \frac{5}{10} = 4 \times \frac{10}{5} = \frac{40}{5} = 8$

e) $\frac{\sqrt{72} \times \sqrt{50}}{-\sqrt{0.04}} = \frac{\sqrt{36 \times 2} \times \sqrt{25 \times 2}}{-\sqrt{\frac{4}{100}}} = \frac{\sqrt{36} \times \sqrt{2} \times \sqrt{25} \times \sqrt{2}}{-\frac{\sqrt{4}}{\sqrt{100}}} = \frac{6 \times \sqrt{2} \times 5 \times \sqrt{2}}{-\frac{2}{10}} = \frac{30 \times \sqrt{2} \times \sqrt{2}}{-\frac{2}{10}} = \frac{30 \times \sqrt{4}}{-\frac{2}{10}} = \frac{30 \times 2}{-\frac{2}{10}} =$
$\frac{60}{-\frac{2}{10}} = 60 \div \left(-\frac{2}{10}\right) = 60 \times \left(-\frac{10}{2}\right) = 60 \times (-5) = -300$

Estimate Square Roots

Estimating square roots can be quite useful, especially when you don't have a calculator handy. Here are some methods to help you estimate square roots:

1. **Using Perfect Squares**

Identify the nearest perfect squares around the number whose square root you want to estimate.

2. **Average Method:**

To get a better estimate, you can use the average of the perfect nearby squares:

$$AVE = \frac{Lower\ Perfect\ Square + Upper\ Perfect\ Square}{2}$$

Then we squared the result:

- If the square of AVE is greater than the number whose square root we want to calculate, we should examine the numbers between the *lower perfect square* and the *AVE* to find the approximate square root value.
- If the square of the *AVE* is less than the number whose square root we want to find, we will examine the numbers between the *upper perfect square* and the *AVE*.

Example

Estimate $\sqrt{20}$.

Solution:

1. The nearest perfect squares are 16 (which is 4^2) and 25 (which is 5^2)
2. To get a better estimate, we can use the average of the perfect squares nearby: Lower perfect square is 4 (because $\sqrt{16} = 4$) and upper perfect square is 5 (because $\sqrt{25} = 5$)

$$AVE = \frac{Lower\ Perfect\ Square + Upper\ Perfect\ Square}{2} = \frac{4+5}{2} = 4.5$$

3. Square the result: $(4.5)^2 = 20.25$

   As the square of AVE is greater than 20, we should examine the numbers between the *lower perfect square* (4) and the *AVE* (4.5) to find the approximate square root value.
4. For convenience, we can write the decimal numbers between 4 and 4.5 up to one decimal place in a table and find the square of all of them. Whichever is closest to 20 is the approximate square root of 20:

Number (x)	4	4.1	4.2	4.3	4.4	4.5
Square of number(x^2)	16	16.81	17.64	18.49	19.8	20.25

Given that 19.8 is the closest number to 20 among those tested, the square root of 20 is approximately 4.4

Word Problems

1. Read the Problem Carefully:

- Understand what the problem is asking.
- Identify the key information and what you need to find.

2. Translate Words to Mathematical Expressions: Look for phrases that indicate operations involving exponents or radicals. For example:

- "Squared" or "raised to the power of 2" indicates an exponent of 2.
- "Square root of" indicates a radical expression.

3. Set Up the Equation:

- Use the information given in the problem to write an equation.
- Assign variables if necessary to represent unknown values.

4. Solve the Equation: Apply the appropriate rules for exponents and radicals:

5. Check Your Solution:

- Verify that your solution makes sense in the context of the problem.
- Substitute your solution back into the original equation to ensure it works.

Example:

A rectangle has a length that is 2^3 times longer than its width. If the area of the rectangle is 256 square units, find the length and width of the rectangle.

Solution:

1. Understand the problem:
   - Length (L) is 2^3 times the width (W).
   - Area (A) is 256 square units.
2. Translate to the equations:
   - Length: $L = 2^3 \times W = 8W$
   - Area: $L \times W = 256$
3. Set up the equation:
   $8W \times W = 256 \rightarrow 8W^2 = 256$
4. Solve the equation:
   $8W^2 = 256 \rightarrow W^2 = \frac{256}{8} = 32$
   $W = \sqrt{32} = \sqrt{16 \times 2} = \sqrt{16} \times \sqrt{2} = 4 \times \sqrt{2} = 4\sqrt{2}$
   $L = 8W = 8 \times 4\sqrt{2} = 32\sqrt{2}$

   So, the width is $4\sqrt{2}$ and the length is $32\sqrt{2}$

www.mathnotion.com

Worksheets

✏Multiplication Property of Exponents
Simplify.
1) $2^4 \times 2^9 \times 2^0$
2) $8^5 \times 2^5$
3) $3^2 \times 15^4 \times 5^2$
4) $8^2 \times 2^6$
5) $(5^3)^2 \times 10^4 \times 2^6$
6) $(2^7 \times 4^5 \times 16^2)^3$
7) $25^3 \times 15^4 \times 3^6$
8) $54^2 \times 36 \times 18^3$
9) $(x^2 \times y^5)^3 \times (x^{7.5} \times y^3)^2$
10) $(a^2 \times b^4 \times a^3 \times b^1)^{3.5}$

✏Division Property of Exponents
Find the quotient.
1) $5^9 \div 5^3$
2) $16^3 \div 4^3$
3) $\frac{2^{10} \div 2^5}{3^5}$
4) $\frac{15^3}{3^3} \times \frac{35^4}{7^4}$
5) $\frac{27^3 \div 9^3}{24^5 \div 24^2}$
6) $\frac{(2^3)^2 \times 8 \times 5^9}{2^2 \times 5^2}$
7) $(3^5 \times 2^5)^2 \div (6^2 \div 6^0)^5$
8) $(\frac{a^{12} \div a^5}{b^3 \times b^4})^2$
9) $((\frac{x^{\frac{5}{2}} \div x^{\frac{1}{2}}}{y^{0.5} \times y^{1.5}})^{-2})^{-1}$
10) $\frac{42^3 \times 14^2}{7^2 \times 21^3} \div \frac{35^3 \times 8}{10^3 \times 7^3}$

✏Negative Exponents and Negative Bases
Solve.
1) $3^{-4} \times 3^6$
2) $-(-5)^4 \times 5^{-10}$
3) $14^{-2} \div (-14)^{-5}$
4) $7^{-5} \times 14^3 \times (-2)^{-9}$
5) $\frac{-30^2 \div 15^2}{(-2)^6}$
6) $\frac{4^{-2}}{15^3} \times \frac{5^{-3}}{3^{-1}} \times \frac{3^2}{2^{-4}}$
7) $(\frac{1}{2})^{-3} \times (4^{-2} \times (-3)^{-7})^{-1}$
8) $\frac{32^{-2} \div (-8)^{-2}}{18^3 \div (-18)^5}$
9) $\frac{80^{-1} \times 45^2}{32^{-2} \times 5^{-1}}$
10) $40^{-2} \times 10^2 \div 8^{-1} \div 3^{-2}$

✏Fraction and Decimal Bases
Evaluate.
1) 0.001^{-3}
2) $(\frac{2}{3})^5 \div (\frac{3}{2})^{-2}$
3) $(\frac{1}{2})^5 \times 0.5^3 \times \frac{6}{12}$
4) $0.75^{-2} \times (\frac{4}{3})^{-2} \times 10^2$
5) $1.2^3 \times (\frac{5}{6})^{-3} \times (\frac{6}{5})^4$
6) $(1\frac{2}{5})^{-4} \div (1.4)^8$
7) $(5^{-1} \times 0.2^3)^{-3} \div 125$
8) $0.002^{-3} \times 2^3 \div 0.1^{-5} \times 7^4$
9) $-\frac{(\frac{2}{3})^{-2} \div 1.5^4 \times 32^{-1}}{-3^{-5}}$
10) $\frac{0.75^2 + 0.75^2 + 0.75^2}{-1.25^2 - 1.25^2 - 1.25^2 - 1.25^2 - 1.25^2}$

✏Scientific Notation
Convert numbers to scientific notation:
1) 10,000,000
2) 2,540,000
3) 0.00156
4) 0.000000852
5) 11.11
6) 21.8452

7) 213.45×10^{-5}
8) $0.2^2 \times 10^3$
9) $1.4 \times 0.002 \times 0.1^{-1}$
10) $12000 \div 10^5 \times 10^{-2}$

Square Roots

Calculate the square roots:

1) $\sqrt{400}$
2) $\sqrt{0.01}$
3) $\sqrt{1.21}$
4) $\sqrt{\frac{16}{81}}$
5) $\sqrt{36} \div \sqrt{49}$
6) $\sqrt{2} \times \sqrt{50}$
7) $\frac{\sqrt{147}}{\sqrt{3}}$
8) $\sqrt{5 \times \sqrt{25 \times 16}}$
9) $\sqrt{175}$
10) $\sqrt{2700}$

Estimate Square Roots

Estimate the square roots of the following numbers to one decimal place:

1) 42
2) 68
3) 95
4) 12
5) 175

Word Problems

Do the word problems:

1) Radioactive substance decays by 25% every year. If the initial amount is 200 grams, how much will remain after 3 years?
2) You invest $2,000 at an annual interest rate of 7%, compounded annually. What will be the value of the investment after 5 years?
3) The radius of a circular garden is $\sqrt{15}$ meters. What is the area of the garden?
4) The length of one side of the square is $10\sqrt{3}$ meters. What is the area and perimeter of this square?
5) If the area of a square is 72 square centimeters, what is the length of one side of the square?
6) The ball is dropped from a height of 100 meters. Each time it hits the ground, it bounces back to 60% of its previous height. How high will the ball bounce after the $3th$ bounce?
7) A car's value depreciates by 20% per year. If the initial value of the car is $25,000, what will its value be after 5 years?
8) A scientist measures the decay of a radioactive substance, which decreases by half every 3 hours. If the initial amount of the substance is 8×10^6 grams, how much of the substance remains after 15 hours?
9) In a right triangle, the lengths of the legs are $\sqrt{50}$ and $\sqrt{72}$. Find the length of the hypotenuse using the Pythagorean theorem.
10) We paint half of a square, and each time we paint half of the remaining white part. After 6 times of painting, what fraction of the square will be painted?

Answer of Worksheets

Multiplication Property of Exponents
1) 2^{13}
2) 2^{20}
3) 15^6
4) 2^{12}
5) 10^{10}
6) 2^{75}
7) 15^{10}
8) 18^7
9) $(xy)^{21}$
10) $(ab)^{17.5}$

Division Property of Exponents
1) 5^6
2) 4^3
3) $(\frac{2}{3})^5$
4) 5^7
5) $\frac{1}{2^9}$
6) 10^7
7) 1
8) $(\frac{a}{b})^{14}$
9) $(\frac{x}{y})^4$
10) 2^5

Negative Exponents and Negative Bases
1) 3^2
2) -5^{-6}
3) -14^3
4) -56^{-2}
5) -2^{-4}
6) 5^{-6}
7) $(-6)^7$
8) $-(\frac{9}{2})^2$
9) 360^2
10) 4.5

Fraction and Decimal Bases
1) 10^9
2) $(\frac{2}{3})^3$
3) $(\frac{1}{2})^9$
4) 10^2
5) 1.2^{10}
6) $(1.4)^{-12}$
7) 5^9
8) 70^4
9) $-(\frac{3}{2})^3$
10) $-(\frac{3}{5})^3$

Scientific Notation
1) 10^7
2) 2.54×10^6
3) 1.56×10^{-3}
4) 8.52×10^{-7}
5) 1.111×10^1
6) 2.18452×10^1
7) 2.1345×10^{-3}
8) 4×10^1
9) 2.8×10^{-2}
10) 1.2×10^{-3}

Square Roots
1) 20
2) 0.1
3) 1.1
4) $\frac{4}{9}$
5) $\frac{6}{7}$
6) 10
7) 7
8) 10
9) $5\sqrt{7}$
10) $30\sqrt{3}$

Estimate Square Roots
1) 6.5
2) 8.2
3) 9.7
4) 3.5
5) 13.2

Word Problems
1) 84.375 grams
2) Approximately $2805.10
3) Approximately 47.12 square meters
4) The area: 300 square meters and the perimeter: $40\sqrt{3}$
5) $6\sqrt{2}$ centimeters
6) 21.6 meters
7) Approximately $8,192
8) 2.5×10^5 grams
9) $\sqrt{122}$
10) $\frac{63}{64}$ of square

www.mathnotion.com

Chapter 7: Algebraic Expressions

Topics that you'll learn in this chapter:
- ✓ Translating Phrases into an Algebraic Statement
- ✓ Identify Terms, Coefficients, and Simplifying
- ✓ Properties of Addition and Multiplication
- ✓ Evaluating Variable Expressions
- ✓ Factor by Distributive Property
- ✓ Factor by Area Model
- ✓ Word Problems
- ✓ Worksheets
- ✓ Answer of Worksheets

Translating Phrases into an Algebraic Statement

Translating phrases into algebraic statements involves understanding the mathematical operations and relationships described in the phrases.

1. **Identify the Unknown(s):** The unknown quantity is typically represented by a variable (e.g., x).

2. **Recognize Key Mathematical Operations:** Look for words that indicate mathematical operations:
   - **Addition**: Sum, plus, increased by, more than, added to.
   - **Subtraction**: Difference, minus, decreased by, less than, subtracted from.
   - **Multiplication**: Product, times, multiplied by.
   - **Division**: Quotient, divided by, per.
   - **Equality**: Equals, is, results in, is the same as.
   - **Inequality**: Less than, greater than, at most, at least.

3. **Translate the Phrase:** Convert each part of the phrase into its corresponding mathematical symbol and arrange them logically.

Examples:

1) Translate the phrase "Three times a number decreased by 7 is 20 "into an algebraic statement.
   Solution:
   - Identify the unknown: Let the unknown number be x.
   - Recognize operations: "Three times" indicates multiplication, "decreased by" indicates subtraction, "is" indicates equality.
   - Translate: $3x - 7 = 20$

2) Translate the phrase "The quotient of a number and 8 is at most 10" into an algebraic statement.
   Solution:
   - Identify the unknown: Let the unknown number be x.
   - Recognize operations: " The quotient " indicates division, "at most " indicates an inequality means less than or equal to.
   - Translate: $\frac{x}{5} \le 10$

Identify Terms, Coefficients, and Simplifying

Identifying Terms and Coefficients:

1. **Terms:** Terms in an algebraic expression are the parts of the expression that are separated by addition (+) or subtraction (−) signs. For example, in the expressions $3x + 4y - 5$, the terms are $3x$, $4y$ and -5.

2. **Coefficients:** A coefficient is a number that multiplies a variable. For example, in the term $3x$, the coefficient is 3 and the variable is x.

Steps to Simplify Algebraic Expressions:

1. **Combine Like Terms:** Like terms are terms that have the same variable raised to the same power. For example, $3xy$ and $5yx$ are like terms, but $3x$ and $3y$ are not.

2. **Add or Subtract Coefficients:** Combine the coefficients of like terms by adding or subtracting them.

Examples:

1) Identify terms and coefficient in following expression:

$-xy + \frac{x}{3} - 4y^2 + 0.5x^3 - 14$

Solution:

- Identifying terms: In this algebraic expression we have 5 different terms:

  $-xy, \frac{x}{3}, -4y^2, 0.5x^3$ and -14

- Identifying coefficients:
  - For the term $-xy$, the coefficient is -1
  - For the term $\frac{x}{3}$, the coefficient is $\frac{1}{3}$
  - For the term $-4y^2$, the coefficient is -4
  - For the term $0.5x^3$, the coefficient is 0.5
  - The term -14 is a constant term and has no variable.

2) Simplify $4y^2 + 2xy - y^2 - 7yx$.

Solution:

Combine like terms: $4y^2 + 2xy - y^2 - 7yx = 4y^2 - y^2 + 2xy - 7xy = (4-1)y^2 + (2-7)xy = 3y^2 - 5xy$

Properties of Addition and Multiplication

Understanding the properties of addition and multiplication is crucial for manipulating and simplifying algebraic expressions. Here is the summary of the key properties provided in a table:

Property	Addition	Multiplication
Commutative	$a + b = b + a$	$a \times b = b \times a$
Associative	$(a + b) + c = a + (b + c)$	$(a \times b) \times c = a \times (b \times c)$
Identity	$a + 0 = a$	$a \times 1 = a$
Distributive	$a(b + c) = ab + ac$	$a(b + c) = ab + ac$
Distributive of multiplication over addition	$(a + b)(c + d) = a(c + d) + b(c + d) = ac + ad + bc + bd$	

☑ The distributive property of multiplication over addition shows that multiplying a sum by another sum can be expected into the sum of individual products.

Examples:

1) Simplify $4(2x + 3) - 2(3x - 1)$.

   Solution:
   - Apply the distributive property: $4(2x + 3) - 2(3x - 1) = 8x + 12 - 6x + 2$
   - Combine like terms: $8x + 12 - 6x + 2 = 8x - 6x + 12 + 2 = (8 - 6)x + 14 = 2x + 14$

2) Simplify $ab(3a - 2b) + 2ab^2 - 4ba^2$.

   Solution:
   - Apply the distributive property: $ab(3a - 2b) + 2ab^2 - 4ba^2 = 3ba^2 - 2ab^2 + 2ab^2 - 4ba^2$
   - Combine like terms: $3ba^2 - 2ab^2 + 2ab^2 - 4ba^2 = 3ba^2 - 4ba^2 - 2ab^2 + 2ab^2 = -ba^2$

3) Simplify $(x - y)(x + y)$.

   Solution:
   - Apply the distributive property:
     $(x - y)(x + y) = x(x + y) - y(x + y) = x^2 + xy - yx - y^2$
   - Combine like terms: $x^2 + xy - yx - y^2 = x^2 + (1 - 1)xy - y^2 = x^2 - y^2$

Evaluating Variable Expressions

Evaluating variable expressions involves substituting the given values for the variables and performing the arithmetic operations. Here's a step-by-step guide to help you through the process:

Steps to Evaluate Variable Expressions

1. **Identify the variable(s) and their values:** Determine the value of each variable in the expression.

2. **Substitute the values into the expression:** Replace each variable with its corresponding value.

3. **Follow the order of operations (PEMDAS):**

    - P: Parentheses/Brackets first
    - E: Exponents (like squares and square roots)
    - MD: Multiplication and Division (from left to right)
    - AS: Addition and Subtraction (from left to right)

Examples

1) Evaluate expression $2x^3 - 3x^2 - x + 2$ using given value: $x = -1$.

    Solution:

    1. Substitute the value into the expression: $2(-1)^3 - 3(-1)^2 - (-1) + 2$
    2. Perform the multiplication then addition: $2 \times (-1) - 3 \times 1 + 1 + 2 = -2$

2) Evaluate expression $2m^2 - 3n + nm$ using given values: $m = -2$ and $n = -3$

    Solution:

    1. Substitute the given values into the expression: $2(-2)^2 - 3(-3) + (-2)(-3)$
    2. Evaluate the exponent: $(-2)^2 = 4$
    3. Perform multiplications:

        - $2(-2)^2 = 2 \times 4 = 8$
        - $-3(-3) = -3 \times (-3) = 9$
        - $(-2)(-3) = (-2) \times (-3) = 6$

    4. Perform the addition: $8 + 9 + 6 = 23$

So, the evaluated value of the $2m^2 - 3n + nm$ is 23.

Factor by Distributive Property

Factoring expressions using the distributive property involves finding common factors in terms of an expression and then rewriting the expression as a product of these common factors. Here's a step-by-step guide to help you understand the process:

Steps to Factor Expressions Using Distributive Property:

1. **Identify Common Factors:** Look for a common factor in each term of the expression.

2. **Factor Out the Greatest Common Factor (GCF):** Rewrite each term as a product of the GCF and another term.

3. **Use the Distributive Property:** Rewrite the expression as the product of the GCF and a binomial.

Examples:

1) Factor $15x - 10$.
   Solution:
   1. Identify the GCF: The GCF of 15 and 10 is 5.
   2. Rewrite the expression: $15x - 10 = 5(3x) - 5(2)$
   3. Factor out the GCF: $= 5(3x - 2)$

   So, the factored form is $15x - 10 = 5(3x - 2)$.

2) Factor $20x^2y^3 + 12y^2x - 4x^3y$.
   Solution:
   1. Identify the GCF: The GCF of $20x^2y^3$, $12y^2x$ and $-4x^3y$ is $4xy$.
   2. Rewrite the expression: $20x^2y^3 + 12y^2x - 4x^3y = 4xy(5xy^2) + 4xy(3y) - 4xy(x^2)$
   3. Factor out the GCF: $= 4xy(5xy^2 + 3y - x^2)$
   4. So, the factored form is: $20x^2y^3 + 12y^2x - 4x^3y = 4xy(5xy^2 + 3y - x^2)$

3) Factor and simplify $\frac{25ab^2 - 10b^2a^2}{15a - 6a^2}$.
   Solution:
   1. Identify the GCF in both numerator and denominator:
      - GCF of $25ab^2$ and $-10b^2a^2$ is $5ab^2$
      - GCF of $15a$ and $-6a^2$ is $3a$
   2. Rewrite the expressions: $\frac{5ab^2(5) - 5ab^2(2a)}{3a(5) - 3a(2a)}$
   3. Factor out the GCF: $\frac{5ab^2(5 - 2a)}{3a(5 - 2a)}$
   4. Simplify: $\frac{25ab^2 - 10b^2a^2}{15a - 6a^2} = \frac{5ab^2(5-2a) \div (5-2a)}{3a(5-2a) \div (5-2a)} = \frac{5ab^2 \div a}{3a \div a} = \frac{5b^2}{3}$.

Factor by Area Model

The area model, also known as the "box method," is a visual way of factoring algebraic expressions. It's particularly helpful for factoring quadratics and other polynomial expressions.

Steps to Use the Area Model:

1. **Draw a Rectangle:** Start by sketching a rectangle and dividing it into parts according to the terms in the expression.
2. **Label the Sections**: Assign a term from the expression to each divided section of the rectangle.
3. **Fill in the Factors:** Write the factors along the sides of the rectangle.
4. **Multiply to Determine Areas:** Calculate the product of the factors to find the area for each section.
5. **Combine the Sections:** Sum the areas of all the sections to obtain the factored expression.

Example:

Factor $18mn^3 - 6m^2n^2 + 12n^3m^4$ using area model.

Solution:

1. Draw a rectangle and label the sections:

$18mn^3$	$-6m^2n^2$	$12n^3m^4$

2. Fill in the Factors: The common factor for $18mn^3, -6m^2n^2$ and $12n^3m^4$ is $6mn^2$ and the other sides of each section are $3n, -m$ and $2nm^3$

	$3n$	$-m$	$2nm^3$
$6mn^2$	$18mn^3$	$-6m^2n^2$	$12n^3m^4$

3. Multiply to Determine Areas:
   - $6mn^2 \times 3n = 18mn^3$
   - $6mn^2 \times (-m) = -6m^2n^2$
   - $6mn^2 \times 2nm^3 = 12n^3m^4$
4. Combine the Sections: Sum the areas of all the sections to obtain the factored expression:

   $18mn^3 - 6m^2n^2 + 12n^3m^4 = 6mn^2(3n - m + 2nm^3)$

Word Problems

Steps to Solve Word Problems about Algebraic Expressions:

1. **Read the Problem Carefully:**
   - Understand what the problem is asking.
   - Identify the key information and any numerical values given.

2. **Define the Variables:**
   - Choose variables to represent the unknown quantities.
   - Clearly state what each variable represents.

3. **Write an Equation:**
   - Translate the words into an algebraic equation using the variables.
   - Use the information given in the problem to set up the equation.

4. **Solve the Equation:**
   - Use algebraic methods to solve the variable(s).
   - Simplify the equation and isolate the variable.

5. **Check the Solution:**
   - Substitute the solution back into the original equation to verify its correctness.
   - Make sure the solution makes sense in the context of the problem.

Example:

Lily and her sister Emma are collecting seashells. Lily has collected 4 times as many seashells as Emma. Together, they collected 125 seashells. How many seashells has each person collected?

Solution:

1. Read the Problem Carefully: Lily has 4 times as many seashells as Emma. Together, they collected 125 seashells.
2. Define the Variables: Let e be the number of seashells Emma has collected. Lily has collected $4e$ seashells.
3. Write an Equation:
   - The total number of seashells is 125.
   - Equation: $e + 4e = 125$
4. Solve the Equation: $e + 4e = 125 \rightarrow 5e = 125 \rightarrow e = \frac{125}{5} = 25$

So, Emma has collected 25 seashells and Lily has collected $4 \times 25 = 100$ seashells.

Worksheets

✏ Translating Phrases into an Algebraic Statement
Write a variable expression for the phrases:
1) Five more than twice a number is 17.
2) The sum of a number and 9 is equal to 3 times the number.
3) Twice the sum of a number and 4 is 18.
4) Half of the opposite of a number is equal to 5.
5) One-third of the difference between twice a number and half of that number is equal to -2.

✏ Identify Terms and Coefficients and Simplifying Algebraic Expressions
Identify terms and coefficients.
1) $-3xy + 2x^2$
2) $\frac{-5a}{3} + 0.5$
3) $1.3y^3 + 16 - x^2$
4) $-5x + 7x + y$
5) $\frac{12m - 8n^2}{4}$

Simplify.
1) $12x - 5y + 7y - 4xy + x - xy$
2) $-1.7a^2 + 2ab + a^2$
3) $\frac{m}{2} - 3n + m + n$
4) $\frac{x^2 + 5xy}{5} - xy - 0.2x^2$
5) $-0.4a^3 + 2a^2 - ab + \frac{a^3}{3} + 8ab - 2a^2$

✏ Properties of Addition and Multiplication
Simplify algebraic expressions using appropriate properties.
1) $-2(a - 4b)$
2) $3(x - 4y) + 2x + 1$
3) $a^2b - 4b + 2a(-ab + 3)$
4) $2x - y(2x + 1) + yx$
5) $2a(a - 3b) + a^2 + ab$
6) $-3(2x - xy) - (2xy + 7x)$
7) $(1 + 2a)(1 + 2a)$
8) $4x(y + 2x) - 5y(x + 1)$
9) $(5m + 1)(2m - 3 + n)$
10) $2(-3x - 4)(1 - x) - 6x^2$

✏ Evaluating Variable Expressions
Evaluate each expression using given values:
1) $5x - y + 2, x = -2$ and $y = 3$
2) $2a^2 + a, a = -1$
3) $-x^2 + xy + 5, x = -1$ and $y = 0$
4) $3m^2 + 4m - 2mn, m = 1$ and $n = -2$
5) $\frac{2a - b}{b + 3}, a = 3$ and $b = -2$
6) $\frac{1 - xy}{y^2 + 1}, x = -3$ and $y = 1$
7) $-z(2x + 3) + 2zx + 3z, x = 1$ and $z = 5$
8) $\frac{4a^3 - 2a}{6a^2 + 2a}, a = 2$
9) $(2x^2 + 1)(2x^2 - 1), x = -4$
10) $\frac{-3m(n^2 + 1)}{4m(-n + m)}, m = -2$ and $n = -1$

Factor Expressions (Using Distributive Property)

Factor following expressions using distributive property:

1) $24x - 16$
2) $5x^2 - 10x + 15$
3) $a^2 - 3a^3$
4) $16x^3 - 4x^2 + 20x^4$
5) $6xy^2 - 18xy + 12x^3y$
6) $5(x-2) + 10x(x-2)$
7) $4xyz^2 - 10x^2yz + 6y^2xz^2$
8) $\frac{3a^2 - 9a}{15a + 12a^2}$ $(a \neq 0, \frac{-5}{4})$
9) $\frac{4xy^3 + 4y^2}{10x^2y + 10x}$ $(x \neq 0)$
10) $\frac{2(a+2) - 10a^2 - 20a}{12a^3 + 24a^2}$ $(a \neq 0, -2)$

Factor by Area Model

Factor following expressions using area model:

1) $2x^4 - 5x^3 + 2x^2$
2) $4ab^2 - ab$
3) $24n^2m^4 - 8n^2$
4) $15x^3y^2 + 10x^4y^3 - 20x^2y^2 + 5x^3y$

Word Problems

Solve the word problems:

1) If the length and width of a rectangle are $5a - 1$ and $3a$, respectively, write the area and perimeter of this rectangle as an algebraic expression.
2) Sarah is twice as old as her brother. Five years ago, she was three times as old as him. How old are they now?
3) The difference between two numbers is 12, and one number is three times the other. Find the numbers.
4) The sum of two consecutive numbers is 45. Find the numbers.
5) A water tank is being filled at a rate of 4 liters per minute. After x minutes, the tank will have a total of 80 liters of water. Write an equation and solve for x.
6) The book has 240 pages. If Sophie reads a fixed number of pages each day and has 30 pages left after one week, how many pages does she read each day?
7) A car travels at a speed of 60 km/h. Another car travels 15 km/h faster and takes 1 hour less to cover the same distance. What is the distance?
8) David has 3 more than twice the number of apples as Tom. Together, they have 24 apples. How many apples does each person have?
9) If the area of a rectangle is $5x^2 - 10x$ and its width is $5x$, calculate its perimeter.
10) A caterpillar grows at a rate of 5 cm per day for the first 10 days, and then at a rate of 2 cm per day for the next x days. If the total growth after $x + 10$ days is 70 cm, write an equation for the caterpillar's growth and solve for x.

www.mathnotion.com

Answer of Worksheets

Translating Phrases into an Algebraic Statement
1) $2x + 5 = 17$.
2) $x + 9 = 3x$
3) $2(x + 4) = 18$.
4) $\frac{-x}{2} = 5$.
5) $\frac{1}{3}\left(2x - \frac{x}{2}\right) = -2$.

Identify Terms and Coefficients and Simplifying Algebraic Expressions
1) Terms: $-3xy$ and $2x^2$, Coefficients: -3 and 2
2) Terms: $\frac{-5a}{3}$ and 0.5, Coefficients: $\frac{-5}{3}$
3) Terms: $1.3y^3, 16$ and $-x^2$, Coefficients: 1.3 and -1
4) $-5x + 7x + y$ Terms: $2x$ and y, Coefficients: 2 and 1
5) $\frac{12m - 8n^2}{4}$ Terms: $3m$ and $-2n^2$, Coefficients: 3 and -2
6) $13x + 2y - 9xy$
7) $-0.7a^2 + 2ab$
8) $\frac{3m}{2} - 2n$
9) 0
10) $\frac{-a^3}{15} + 7ab$

Properties of Addition and Multiplication
1) $-2a + 8b$
2) $-12y + 5x + 1$
3) $-a^2b - 4b + 6a$
4) $2x - y - yx$
5) $3a^2 - 5ab$
6) $-13x + xy$
7) $1 + 4a + 4a^2$
8) $8x^2 - xy - 5y$
9) $10m^2 - 13m + 5mn + n - 3$
10) $2x - 8$

Evaluating Variable Expressions
1) -11
2) 1
3) 4
4) 11
5) 8
6) 2
7) 0
8) 1
9) $1,023$
10) $\frac{3}{2}$

Factor Expressions (Using Distributive Property)
1) $8(3x - 2)$
2) $5(x^2 - 2x + 3)$
3) $a^2(1 - 3a)$
4) $4x^2(4x - 1 + 5x^2)$
5) $6xy(y - 3 + 2x^2)$
6) $(x - 2)(5 + 10x)$
7) $2xyz(2z - 5x + 3yz)$
8) $\frac{(a-3)}{(5+4a)}$
9) $\frac{2y^2}{5x}$
10) $\frac{1-5a}{6a^2}$

Factor by Area Model
1) $x^2(2x^2 - 5x + 2)$

	$2x^2$	$-5x$	2
x^2	$2x^4$	$-5x^3$	$2x^2$

2) $ab(4b-1)$

	$4b$	-1
ab	$4ab^2$	$-ab$

3) $8n^2(3m^4-1)$

	$3m^4$	-1
$8n^2$	$24n^2m^4$	$-8n^2$

4) $5x^2y(3xy+10x^2y^2-4y+x)$

	$3xy$	$2x^2y^2$	$-4y$	x
$5x^2y$	$15x^3y^2$	$10x^4y^3$	$-20x^2y^2$	$5x^3y$

Word Problems
1) $Area: 15a^2-3a, Perimeter: 16a-2$
2) Sarah is 20 years old and his brother is 10 years old
3) The smaller number is 6 and the larger number is 18
4) The two consecutive numbers are 22 and 23
5) $4x=80$ and $x=20$
6) 30 pages
7) $300\ km$
8) Tom has 7 apples and David has 17 apples
9) $12x-4$
10) $50+2x=70$ and $x=10$

Chapter 8: Equations and Inequalities

Topics that you'll learn in this chapter:

- ✓ One-Step Equations
- ✓ Multi-Step Equations
- ✓ One-Step Inequalities
- ✓ Multi-Step Inequalities
- ✓ Graphing Inequalities
- ✓ Word Problems
- ✓ Worksheets
- ✓ Answer of Worksheets

One-Step Equations

One-step equations are algebraic equations that can be solved in a single step by performing the inverse operation to isolate the variable. The goal is to find the value of the variable that makes the equation true. These equations typically involve only one operation: addition, subtraction, multiplication, or division.

Types of One-Step Equations

1. **Addition Equations**: $x + a = b$
2. **Subtraction Equations**: $x - a = b$
3. **Multiplication Equations**: $ax = b$
4. **Division Equations**: $\frac{x}{a} = b$

Steps to Solve One-Step Equations

1. **Identify the operation being performed on the variable**: This could be addition, subtraction, multiplication, or division.
2. **Perform the inverse operation**: Apply the inverse operation to both sides of the equation to isolate the variable.

Examples:

1) Solve $x - 8 = 5$.
   Solution:
   1. Identify the operation: Subtraction of 8.
   2. Perform the inverse operation: Add 8 to both sides.
      $x - 8 + 8 = 5 + 8$
      $x = 13$

2) Solve $-5x = 12$.
   Solution:
   1. Identify the operation: Multiplication by -5.
   2. Perform the inverse operation: Divide both sides by -5.
      $\frac{-5x}{-5} = \frac{12}{-5}$
      $x = -\frac{12}{5}$

3) Solve $\frac{2x}{3} = 8$.
   Solution:
   3. Identify the operation: Division by $\frac{2}{3}$.
   4. Perform the inverse operation: Multiply both sides by $\frac{2}{3}$.
      $\frac{2x}{3} \times \frac{3}{2} = 8 \times \frac{3}{2} \rightarrow x = \frac{24}{2} = 12.$

Multi-Step Equations

Multi-step equations are algebraic equations that require more than one operation to isolate the variable and solve the equation. These equations often involve a combination of addition, subtraction, multiplication, division, and sometimes parentheses or fractions. Solving multi-step equations involves performing a series of steps in a systematic manner to isolate the variable.

Characteristics of Multi-Step Equations

- **Multiple Operations**: Involves more than one mathematical operation.
- **Combining Like Terms**: May require combining like terms on one or both sides of the equation.
- **Distributive Property**: May need to use the distributive property to remove parentheses.
- **Fractions or Decimals**: May include fractions or decimals that need to be cleared.

Steps to Solve Multi-Step Equations

1. **Simplify both sides of the equation** (if needed): Combine like terms and distribute any factors.
2. **Move variables to one side**: Use addition or subtraction to get all the variables on one side of the equation.
3. **Isolate the variable**: Use inverse operations (addition, subtraction, multiplication, division) to solve for the variable.
4. **Check your solution**: Substitute the solution back into the original equation to ensure it works.

Examples:

1) Solve $2(x - 3) = 4x + 6$.
   Solution:
   1. Distribute on the left side: $2x - 6 = 4x + 6$
   2. Move variables to one side: Subtract $2x$ from both sides:
      $2x - 2x - 6 = 4x - 2x + 6$
      $-6 = 2x + 6$
   3. Isolate the variable: subtract 6 from both sides:
      $-6 - 6 = 2x + 6 - 6$
      $-12 = 2x$
   4. Divide both sides by 2: $\frac{-12}{2} = \frac{2x}{2} \rightarrow x = -6$
   5. Check your solution: substitute $x = -6$ into both sides of the original equations:
      $2(x - 3) = 2(-6 - 3) = -18$
      $4x + 6 = 4(-6) + 6 = -18$

One-Step Inequalities

One-step inequalities are algebraic expressions that can be solved in a single step to find the range of values that make the inequality true. These inequalities involve performing one mathematical operation to isolate the variable. The solution to a one-step inequality is typically represented on a number line, showing all possible values that satisfy the inequality.

Types of One-Step Inequalities

1. **Addition Inequality**: $x + a < b$
2. **Subtraction Inequality**: $x - a > b$
3. **Multiplication Inequality**: $ax \leq b$
4. **Division Inequality**: $\frac{x}{a} \geq b$

Steps to Solve One-Step Inequalities

1. **Identify the operation performed on the variable**: Addition, subtraction, multiplication, or division.

2. **Perform the inverse operation**: Apply the inverse operation to both sides of the inequality to isolate the variable.

☑ Remember to flip the inequality sign when multiplying or dividing both sides by a negative number (e.g., if multiplying by -1, $<$ becomes $>$).

Examples:

1) Solve $x + 8 < -5$.

   Solution:

   1. Identify the operation: addition to 8
   2. Perform the inverse operation: Subtract 8 from both sides:

      $x + 8 - 8 < -5 - 8$

      $x < -13$

2) Solve $\frac{x}{-3} \leq 1$.

   Solution:

   1. Identify the operation: Division by -3
   2. Perform the inverse operation: Multiply both sides by -3 and flip the ($\leq$) into ($\geq$):

      $\frac{x}{-3} \times (-3) \leq 1 \times (-3)$

      $x \geq -3$

www.mathnotion.com

Multi-Step Inequalities

Multi-step inequalities are inequalities that require more than one operation to isolate the variable and solve for the range of values that make the inequality true. These steps can include a combination of addition, subtraction, multiplication, division, and sometimes the use of parentheses or the distributive property.

Steps to Solve Multi-Step Inequalities

1. **Simplify each side of the inequality if necessary:**

   - Combine like terms.
   - Distribute any factors.

2. **Move all variable terms to one side of the inequality:** Use addition or subtraction to get all variable terms on one side and constant terms on the other side.

3. **Isolate the variable:**

   - Use addition, subtraction, multiplication, or division to solve the variable.
   - **Important:** If you multiply or divide by a negative number, you must reverse the inequality sign.

Examples:

1) Solve $\frac{x}{-2} + 2 < 1$

   Solution:
   1. Move constant terms to one side: Add -2 to both sides:
      $\frac{x}{-2} + 2 - 2 < 1 - 2$
      $\frac{x}{-2} < -1$
   2. Isolate the variable: multiply both sides by -2 and reverse the direction of the inequality sign.
      $\frac{x}{-2} \times (-2) < -1 \times (-2) \to x > 2$

2) Solve the inequality $3(x - 2) \leq 2x + 4$

   Solution:
   1. Distribute and simplify: $3x - 6 \leq 2x + 4$
   2. Move variable terms to one side: Subtract $2x$ from both sides:
      $3x - 6 - 2x \leq 2x + 4 - 2x$
      $x - 6 \leq 4$
   3. Isolate the variable: Add 6 to both sides:
      $x - 6 + 6 \leq 4 + 6$
      $x \leq 10$

Graphing Inequalities

When graphing one-variable inequalities, we typically work with a number line rather than a coordinate plane. Here's a step-by-step guide:

Steps to Graph One-Variable Inequalities

1. **Identify the inequality:** Common forms include $x > a, x \geq a, x < a$ and $x \leq a$.
2. **Draw a number line:** Mark the key value (a) on the number line.
3. **Determine the type of circle to use at the key value:**
   - Use an **open circle** for $>$ or $<$ (indicating the value is not included).
   - Use a **closed circle** for $\geq$ or $\leq$ (indicating the value is included).
4. **Shade the appropriate region:**
   - For $x > a$ or $x \geq a$, shade to the right of a.
   - For $x < a$ or $x \leq a$, shade to the left of a.

Examples:

1) Solve $-\frac{1}{3}x + 2x \leq 5$ and then graph the solution.

   Solution:
   1. Simplify the left side:
      $$-\frac{1}{3}x + \frac{6}{3}x \leq 5$$
      $$\frac{5}{6}x \leq 5$$
   2. Isolate the variable: Multiply both sides by $\frac{6}{5}$:
      $$\frac{5}{6}x \times \frac{6}{5} \leq 5 \times \frac{6}{5}$$
      $$x \leq \frac{30}{5} \rightarrow x \leq 6$$
   3. Draw a number line and use a closed circle:

   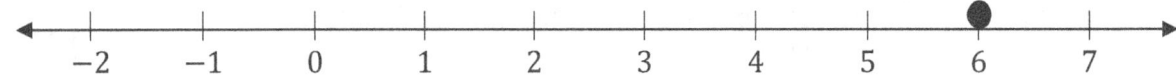

   4. Shade the appropriate region:
   5.

   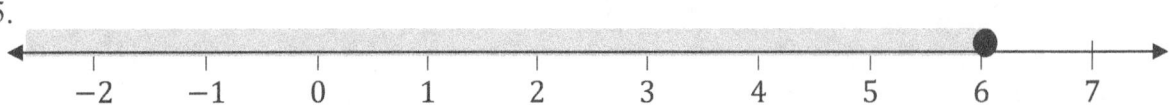

Word Problems

Steps to Solve Word Problems

1. **Understand the Problem:**
   - Read the problem carefully to understand what is being asked.
   - Identify the key pieces of information and the unknown variable.

2. **Define the Variable:** Assign a variable (often x or another letter) to represent the unknown quantity.

3. **Translate the Words into an Equation or Inequality:** Use the information given in the problem to write an equation or inequality that represents the situation.

4. **Solve the Equation or Inequality:** Use algebraic methods to isolate the variable and find its value.

5. **Check the Solution:**
   - Substitute the solution back into the original problem to ensure it makes sense.
   - Verify, the solution meets all the conditions stated in the problem.

Example:

1) A school is organizing a field trip, and the cost of the trip is covered by contributions from students. The total cost of the trip is $720. Each student contributes $10, but 5 students could not make it, so each of the remaining students had to contribute an additional $2 to cover the total cost. How many students were originally supposed to go on the trip?

Solution:
1. Understand the problem: We need to find the original number of students.
2. Define the variable: Let x represent the original number of students.
3. Translate the words into an equation:
   - The total contribution if all students participated: $10x$.
   - Since 5 students didn't make it, there were $x - 5$ students who contributed an additional $2 each, so each of these students paid $12.
   - The total amount collected: $12(x - 5)$.
4. Set up the equation:
   $12(x - 5) = 720$
   $12x - 60 = 720$
   $12x = 60 + 720$
   $12x = 780$
   $x = \frac{780}{12} = 65$

So, 65 students were supposed to go on the trip.

www.mathnotion.com

Worksheets

🍃 One-Step Equations

Solve the one-step equations:
1) $x - 6 = 12$
2) $8 + x = -19$
3) $0.2 = 1.5 - x$
4) $\frac{5}{3}x = -2$
5) $\frac{x}{-4} = 0.25$
6) $1.8x = -4.8$
7) $\frac{1}{5} = -1\frac{3}{10} + x$
8) $0.6 = -x - \frac{7}{8}$
9) $\frac{1}{9} = \frac{-x}{5}$
10) $\frac{x}{0.5} = -\frac{2}{5}$

🍃 Multi-Step Equations

Solve the multi-step equations:
1) $5x - 3 = 7$
2) $-2 + 4x = -8$
3) $\frac{2x}{5} - 2 = \frac{1}{3}$
4) $3(2x + 7) = 16x$
5) $-0.1(10 - 30x) = 2(x + 4)$
6) $\frac{6x-11}{2} = \frac{-1}{3}$
7) $\frac{x}{10} + \frac{-2}{5} = \frac{2x-1}{6}$
8) $\frac{-3x+4}{3} = -\frac{2x+5}{6}$
9) $1.5(3x - 4) = -2x + 2(0.5x - \frac{1}{2})$
10) $\frac{-3}{4}(x - 2x + 5) = \frac{-2x+5}{8}$

🍃 One-Step Inequalities

Solve the one-step inequalities:
1) $x - 5 < -7$
2) $2x > -3$
3) $12 \geq 6 + x$
4) $\frac{-x}{6} > -2$
5) $\frac{3}{4} - x \leq 8$
6) $\frac{x}{4} < -0.3$
7) $\frac{1}{2} > x + \frac{5}{6}$
8) $-4 \leq \frac{-x}{2.5}$
9) $x - 1.23 > -7.89$
10) $1.5 + x \geq \frac{-3}{5}$

🍃 Multi-Step Inequalities

Solve the multi-step inequalities:
1) $\frac{x}{4} + 4 < 3$
2) $5(x - 3) > 2x - 1$
3) $-x + 2 + 4x \geq -5x$
4) $3x + 4(-x + 0.2) \leq 0.5$
5) $-2(0.2x + 2) > -3(x + 3)$
6) $\frac{6x-2}{5} < \frac{-x}{5} + 1$
7) $4.8 - 2.5x + \frac{5}{10} \geq \frac{-x}{10}$
8) $\frac{-2x+8}{5} \leq \frac{3x+2}{20} + 1.2$
9) $\frac{2}{3}(-6x + 12) > \frac{-3}{5}(-10x + 25)$
10) $\frac{2.6x+1.2}{0.2} < 5(0.1x + \frac{-2}{10})$

🚩Graphing Inequalities

Solve and graph them on the number line.

1) $-2x + 3 < 4x$
2) $1 + \frac{-x}{5} > -2$
3) $-3(x + 5) \geq 2x$
4) $\frac{2x}{5} - \frac{1}{10} \geq x$
5) $-(4 - 3x) \geq 2x + 5$

🚩Word Problems

Solve.

1) Sarah runs around a track that is 450 meters long. She runs 6 laps more than John, and together they run 4,500 meters. How many laps do each run?
2) The sum of three consecutive even numbers is 102. What is the middle number?
3) The temperature outside is $6°C$. If the temperature decreases by $4°C$ per hour, how many hours will it take for the temperature to reach $-2°C$?
4) A rectangular garden has a perimeter of 60 meters. The length is 2 meters more than three times the width. What are the dimensions of the garden?
5) Tim receives $10 each week for his allowance. He wants to save at least $100 for a new game. How many weeks will it take him to save enough?
6) A water tank fills at a rate of 6 liters per minute. After x minutes, the tank has 120 liters of water. Write an equation to represent this and solve for x.
7) A swimming pool can be filled in 8 hours using two pipes. The first pipe can fill the pool in 12 hours on its own. How long does it take for the second pipe to fill the pool on its own?
8) A gardener is planting flowers in two rows. The first row has 4 more flowers than the second row. If the gardener plants a total of 68 flowers in both rows, how many flowers are planted in each row?
9) A bookstore sold 150 books. Some were sold at $10 each, and others at $15 each. The total revenue was $1800. How many books were sold at each price?
10) A phone's battery drains at a rate of 5% per hour. If the phone starts with 100% battery, how many hours will the battery be at most 25%?

Answer of Worksheets

One-Step Equations
1) $x = 18$
2) $x = -27$
3) $x = 1.3$
4) $x = -\frac{6}{5}$
5) $x = -1$
6) $x = -\frac{8}{3}$
7) $x = \frac{3}{2}$
8) $x = \frac{-59}{40}$
9) $x = \frac{-5}{9}$
10) $x = -\frac{1}{5}$

Multi-Step Equations
1) $x = 2$
2) $x = -1.5$
3) $x = \frac{35}{6}$
4) $x = 2.1$
5) $x = 9$
6) $x = \frac{31}{18}$
7) $x = -1$
8) $x = \frac{13}{4}$
9) $x = \frac{10}{11}$
10) $x = \frac{35}{8}$

One-Step Inequalities
1) $x < -2$
2) $x > -1.5$
3) $x \leq 6$
4) $x < 12$
5) $x \geq -\frac{29}{4}$
6) $x < -1.2$
7) $x < \frac{-1}{3}$
8) $x \leq 10$
9) $x > -6.66$
10) $x \geq -2.1$

Multi-Step Inequalities
1) $x < -4$
2) $x > \frac{14}{3}$
3) $x \geq \frac{-1}{4}$
4) $x \geq 0.3$
5) $x > -\frac{25}{13}$
6) $x < 1$
7) $x \leq \frac{53}{24}$
8) $x \geq \frac{6}{11}$
9) $x < 2.3$
10) $x < -0.56$

Graphing Inequalities

1) $x > \frac{1}{2}$

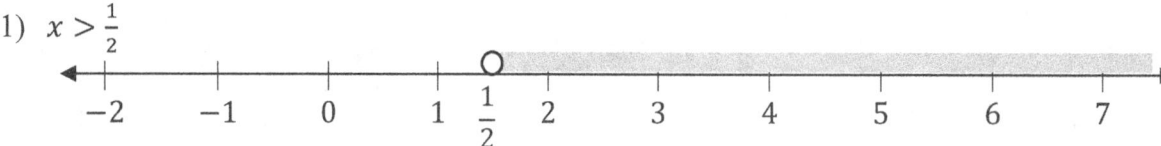

2) $x < 15$

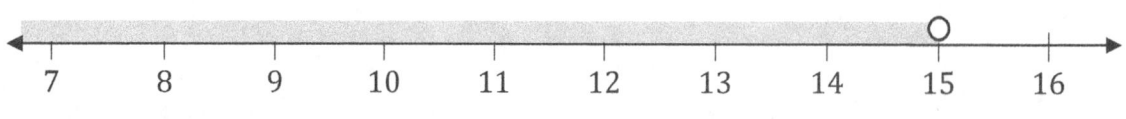

3) $x \leq -3$

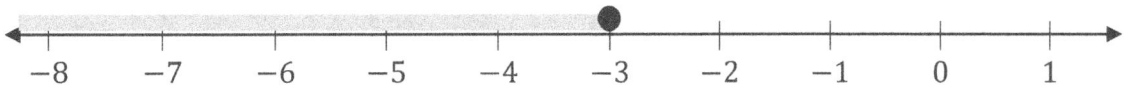

4) $x \leq \frac{-1}{6}$

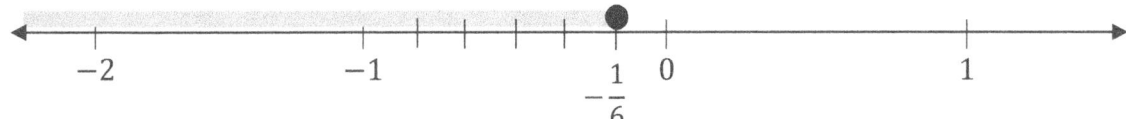

5) $x \geq 9$

Word Problems
1) Joun runs 2 laps and Sarah runs 8 laps
2) The middle number is 34
3) 2 hours
4) Width: 7 meters and Length: 23 meters
5) At least 10 weeks
6) $6x = 120$ and $x = 20$
7) 24 hours
8) 32 flowers in the second row and 36 flowers in the first row
9) 90 books at $10 each and 60 books at $15 each
10) After at least 15 hours

Chapter 9: Linear Functions

Topics that you'll learn in this chapter:

- ✓ Slope from Graph
- ✓ Slope from Two Points
- ✓ Identifying Slope and Y-Intercept from
- ✓ Rate of Change
- ✓ Graphing Lines Using Line Equation
- ✓ Writing Linear Equations
- ✓ Write an Equation from Graph
- ✓ Graphing Linear Inequalities
- ✓ System of Equations
- ✓ Point Coordinate and Finding Midpoint
- ✓ Finding Distance of Two Points
- ✓ Word Problems
- ✓ Worksheets

Slope from Graph

What is slope: The slope is like a "tilt" of a line. It tells us how steep the line is.

How to Calculate: If you have two points on a line, you can find the slope by figuring out how much you go up or down (change in y) and how much you go left or right (change in x). The slope is the change in y divided by the change in x.

$Slope = \frac{change\ of\ y}{change\ of\ x}$

Example:

Find the slope of the line:

Solution:

If we move from point (0,1) to (3,3):

Change in y: $3 - 1 = 2$

Change in x: $3 - 0 = 3$

Slope: $\frac{2}{3}$

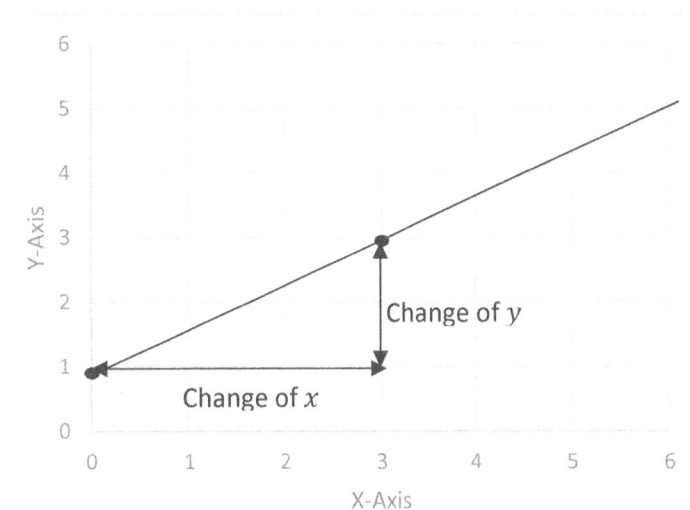

Slope from Two Points

Step-by-Step Guide to Finding the Slope

1. **Identify the coordinates of the two points:** Let's say the points are (x_1, y_1) and (x_2, y_2).

2. **Use the slope formula:** The formula to find the slope (m) between two points is:

$$m = \frac{y_2 - y_1}{x_2 - x_1}$$

This formula calculates the change in y (vertical change) divided by the change in x (horizontal change).

Example:

Find the slope between the points (2,3) and (5,11).

Solution:
1. Identify the coordinates:
   - $(x_1, y_1) = (2,3)$
   - $(x_2, y_2) = (5,11)$
2. Plug the coordinates into the slope formula: $m = \frac{11-3}{5-2} = \frac{8}{3}$

www.mathnotion.com

Identifying Slope and Y-Intercept

What is a Linear Equation?
- A linear equation is an equation that forms a straight line when graphed on a coordinate plane.
- It can be written in the form $y = mx + b$, where:
  - y is the dependent variable (the value you get out).
  - x is the independent variable (the value you put in).
  - m is the slope (how steep the line is).
  - b is the y-intercept (where the line crosses the y-axis).

Steps to Identify Slope and y-Intercept:

- **Write the Equation in Slope-Intercept Form:** Make sure your equation looks like $y = mx + b$
- **Identify the Slope (m):** The coefficient of x is the slope. It's the number in front of x.
- **Identify the y-Intercept (b):** The constant term (the number without x) is the y-intercept. It's where the line crosses the y-axis.

Examples:

1) Identify slope and y-intercept of $y = -5x + 4$.
   Solution:
   1. Write the equation in slope-intercept form: The equation is already in the form $y = mx + b$
   2. Identify the slope (m): The slope (m) is -5.
   3. Identify the y-intercept (b): The y-intercept is 4

2) Identify the slope and y-intercept of $2y - 4x = 8$.
   Solution:
   1. Rearrange the equation into slope-intercept form:
      Start by solving for y: $2y - 4x = 8 \rightarrow y = 2x + 4$
   2. Identify the slope (m): The slope (m) is 2
   3. Identify the y-intercept (b): The y-intercept is 4

Rate of Change

The "rate of change" is a fundamental concept in mathematics, particularly important for understanding how one quantity changes in relation to another.

Definition of Rate of Change: It measures how a quantity changes over a specific period or interval. It can be thought of as the "speed" at which one variable changes in relation to another.

Formula for Rate of Change:

$$Rate\ of\ Change = \frac{Change\ in\ Dependent\ Variable}{Change\ in\ Independent\ Variable}$$

Average Rate of Change: The average rate of change of a function over an interval, measures how much the function's value changes on average between two points. It is similar to the concept of slope for linear functions and can be thought of as the slope of the secant line connecting two points on the function's graph.

Formula for the Average Rate of Change:

$$Average\ Rate\ of\ Change = \frac{f(b) - f(a)}{b - a}$$

Here:

- $f(a)$ and $f(b)$ are the values of the function at points a and b respectively.
- a and b are the points in the interval over which you want to find the average rate of change.

Examples:

1) The temperature in a city increases from $15°C$ at $6\ AM$ to $25°C$ at $2\ PM$. Calculate the average rate of temperature change in degrees Celsius per hour ($°C/h$).
   Solution:
   - Calculate the change in temperature: $T_2 - T_1 = 25°C - 15°C = 10°C$
   - Calculate the change in time: $t_2 - t_1 = 2\ PM - 6\ AM = 8$ hours.
   - Use the formula for average rate of change: $Average\ Rate\ of\ Change = \frac{T_2 - T_1}{t_2 - t_1} = \frac{10°C}{8\ hours} = 1.25°C/h$

2) What is the average rate of change of the function: $f(x) = 12x^3 - x^2$ from $x = -1$ to $x = 1$?
   Solution:
   - Evaluate $f(-1)$: $f(-1) = 12(-1)^3 - (-1)^2 = 12 \times (-1) - 1 = -12 - 1 = -13$
   - Evaluate $f(1)$: $f(1) = 12\ (1)^3 - (1)^2 = 12 \times 1 - 1 = 11$
   - Calculate the average rate of change:

$$Average\ Rate\ of\ Change = \frac{f(1) - f(-1)}{1 - (-1)} = \frac{11 - (-13)}{2} = \frac{24}{2} = 12$$

www.mathnotion.com

Graphing Lines Using Line Equation

Steps to Graph a Line

1. **Identify the Line Equation:**

    - The equation of the line is typically given in the slope-intercept form: $y = mx + b$
    - m is the slope of the line.
    - b is the y-intercept, where the line crosses the y-axis.

2. **Plot the Y-Intercept:** Start by plotting the y-intercept (b) on the y-axis. This is your first point.

3. **Use the Slope to Find Another Point:**

    - From the y-intercept, use the slope (m) to find another point.
    - If the slope is positive, move up (rise) and to the right (run). If negative, move down and to the right.

☑ In general, by assigning different values to x and obtaining the corresponding values of y, various points can be found. By connecting these points together, the line can be drawn.

Example

Graph the line with the equation $y = 2x + 3$.

Solution:

1. Identify the line equation:
    - Slope (m): 2
    - y-intercept (b): 3
2. Plot the y-intercept: Start by plotting the point (0,3) on the y-axis.
3. Use the slope to find another point:
    - Slope: $\frac{2}{1}$ (rise of 2, run of 1)
    - Form (0,3), move up 2 units and 1 unit to the right.
    - Plot the next point at (1,5).

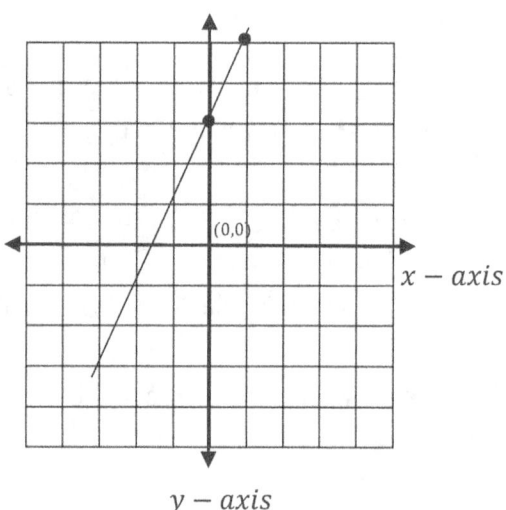

Writing Linear Equations

Writing linear equations involves determining the relationship between two variables, typically x and y, that forms a straight line when plotted on a graph. Linear equations can be written in various forms, including slope-intercept form, point-slope form, and standard form.

1. **Slope-Intercept Form** $(y = mx + b)$: Useful when you know the slope and y-intercept.
2. **Point-Slope Form** $(y - y_1 = m(x - x_1))$: Useful when you know the slope and a point on the line.
3. **Standard Form** $(Ax + By = C)$: Often used in algebra problems and easier to work with for certain types of calculations.

Steps to Write a Linear Equation:

1. **Identify the slope (m):** The slope tells you how much y increases or decreases when x increases by 1. Let's say the line passes through (x_1, y_1) and (x_2, y_2) so the slope will calculate by this formula:
$$m = \frac{y_2 - y_1}{x_2 - x_1}$$
2. **Identify the y-intercept (b):** The y-intercept is where the line crosses the y-axis (the point where $x = 0$).
3. **Write the equation:** Based on the information provided about the line, apply one of the linear equation forms.

Examples:

1) If the slope of a line is -3 and it passes through the point $(1,4)$, find the equation of this line.
   Solution:
   1. Find the slope (m): Here, $m = -3$.
   2. Find the y-intercept (b): Substitute the slope and given point into the equation: Substitute $x = 1$, $y = 4$ and $m = -3$ into the equation to find b:
   $4 = -3(1) + b \rightarrow 4 = -3 + b \rightarrow b = 7$
   3. Write the equation: Now that you have $m = -3$ and $b = 7$, the equation is: $y = -3x + 7$

2) Write a linear equation that passes through $(2,4)$ and $(1, -2)$.
   Solution:
   1. Find the slope (m):
   $m = \frac{y_2 - y_1}{x_2 - x_1} = \frac{-2-4}{1-2} = \frac{-6}{-1} = 6$
   2. Apply the Point-Slope form by using the slope and one of the points (we use $(2,4)$):
   $y - y_1 = m(x - x_1) \rightarrow y - 4 = 6(x - 2)$
   $y - 4 = 6x - 12$
   $y = 6x - 8$
   So, the equation is $y = 6x - 8$

Write an Equation from Graph

Writing a linear equation from a graph involves identifying the key components of the line: the slope and the y-intercept.

Steps to Write a Linear Equation from a Graph:

1. **Identify Two Points on the Line**: Select two clear points on the line, preferably where the line crosses the grid lines for accuracy. Let's call these points (x_1, y_1) and (x_2, y_2).

2. **Calculate the Slope:** The slope of the line is the "rise" over the "run," which is the change in y over the change in x:

$$m = \frac{y_2 - y_1}{x_2 - x_1}$$

3. **Find the Y-intercept (b):** The y-intercept is the point where the line crosses the y-axis ($x = 0$). If it's clearly shown on the graph, you can read it directly. Otherwise, use one of the points and the slope to find it.

4. **Use the Slope-Intercept Form:** Substitute the values of m (slope) and b (y-intercept) into the slope-intercept form of linear equation ($y = mx + b$).

Example:

Write the equation of the graph.

Solution:

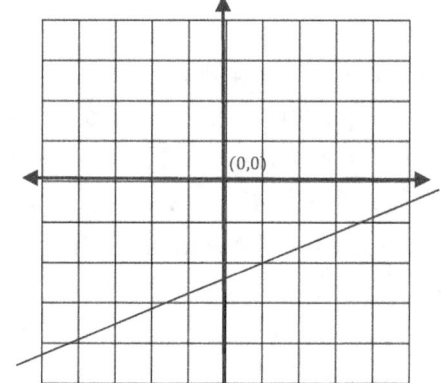

1. Identify two points: This line passes through the points $(1, -2)$ and $(-4, -4)$

2. Calculate the slope (m):
$m = \frac{-4-(-2)}{-4-1} = \frac{-2}{-5} = \frac{2}{5}$

3. Find the y-intercept (b): Use the slope-intercept form with one of the points, say $(1, -2)$:

$-2 = \frac{2}{5} \times 1 + b \rightarrow -2 = \frac{2}{5} + b \rightarrow b = -2 - \frac{2}{5} = \frac{-12}{5}$

4. Write the equation: $y = \frac{2}{5}x - \frac{12}{5}$

Graphing Linear Inequalities

Graphing linear inequalities involves shading a region of the coordinate plane that satisfies inequality. This is different from graphing linear equations, where you only plot a line.

Steps to Graph Linear Inequalities

1. **Rewrite the inequality in slope-intercept form $y = mx + b$** (if it's not already).

2. **Graph the corresponding linear equation $y = mx + b$ as if it was the equation:**
   - Use a solid line if the inequality is $\leq$ or $\geq$.
   - Use a dashed line if the inequality is $<$ or $>$.

3. **Determine which side of the line to shade**:
   - Pick a test point not on the line (often (0,0) is convenient unless it's on the line).
   - Substitute the test point into the inequality.
   - If the inequality is true, shade the side containing the test point.
   - If the inequality is false, shade the opposite side.

Example:

Graph $y > -\frac{1}{2}x + 1$.

Solution:

1. The inequality is already in slope-intercept form.
2. Graph the line $y = -\frac{1}{2}x + 1$ using a dashed line because the inequality is $>$.
3. Pick the test point (0,0):
   - Substitute into the inequality: $0 > -\frac{1}{2}(0) + 1$ which simplifies to $0 > 1$. This is false.
   - Since the inequality is false for (0,0), shade the region above the line.

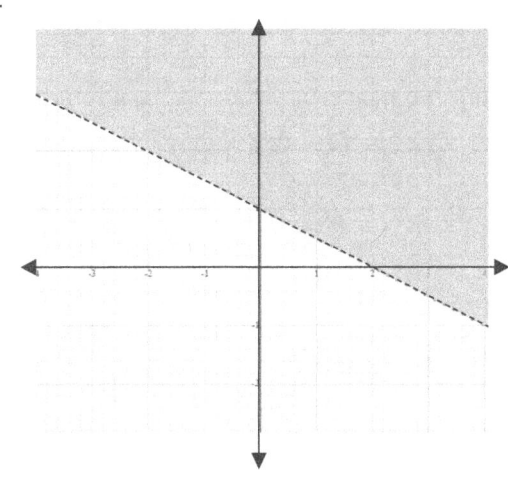

System of Equations

A system of equations is a set of two or more equations with the same set of variables. The goal is to find the values of the variables that satisfy all the equations in the system simultaneously.

Solving a System of Equations:

There are several methods to solve a system of equations:

1. **Graphing:**
   - Plot each equation on the same coordinate plane.
   - The point where the lines intersect is the solution.

2. **Substitution:**
   - Solve one equation for one variable.
   - Substitute that expression into the other equation.
   - Solve for the remaining variable.

3. **Elimination (Addition/Subtraction):**
   - Add or subtract the equations to eliminate one of the variables.
   - Solve for the remaining variable.

Example

Solve the system of equations (using substitution method):

$$\begin{cases} x + y = 7 \\ 2x - y = 3 \end{cases}$$

Solution:

1. Solve the first equation for y:
   $y = 7 - x$
2. Substitute y into the second equation:
   $2x - (7 - x) = 3$
   $2x - 7 + x = 3$
   $3x - 7 = 3$
   $3x = 10$
   $x = \frac{10}{3}$
3. Substitute x back into $y = 7 - x$:
   $y = 7 - \frac{10}{3} = \frac{21}{3} - \frac{10}{3} = \frac{11}{3}$

   So, the solution is $x = \frac{10}{3}$ and $y = \frac{11}{3}$.

Point Coordinate and Finding Midpoint

Point Coordinates:

- **Definition:** A point in a coordinate plane is defined by an ordered pair of numbers (x, y).

- **Coordinate Plane:** A grid formed by a horizontal line (x-axis) and a vertical line (y-axis). The intersection of these two lines is called the origin, denoted by $(0,0)$.

- **Example:** If a point is located at $(3,4)$, it means it's 3 units to the right of the origin and 4 units up.

Finding the Midpoint:

The midpoint of a line segment connecting two points is the point that is exactly halfway between them.

- **Formula:** The midpoint M of a line segment connecting points (x_1, y_1) and (x_2, y_2) is given by:

$$M = (\frac{x_1 + x_2}{2}, \frac{y_1 + y_2}{2})$$

- **Steps:**
    1. Identify the coordinates of the endpoints.
    2. Apply the midpoint formula.
    3. Substitute the coordinates into the formula.
    4. Calculate the averages.

Example:

Find the midpoint of the line segment connecting $(2,3)$ and $(4,7)$.

Solution:

1. Identify the coordinate:
    - $(x_1, y_1) = (2,3)$
    - $(x_2, y_2) = (4,7)$

2. Apply the midpoint formula and substitute the coordinates into the formula:

    $M = \left(\frac{x_1+x_2}{2}, \frac{y_1+y_2}{2}\right) = \left(\frac{2+4}{2}, \frac{3+7}{2}\right)$

3. Calculate the average:

    $M = \left(\frac{6}{2}, \frac{10}{2}\right) = (3,5)$

Finding Distance of Two Points

Finding the distance between two points in a coordinate plane involves using the distance formula, which is derived from the Pythagorean theorem.

Distance Formula:

The distance d between two points (x_1, y_1) and (x_2, y_2) is given by:

$$d = \sqrt{(x_2 - x_1)^2 + (y_2 - y_1)^2}$$

Step-by-Step Guide

1. **Identify the coordinates of the two points**:
   - Let's say the two points are (x_1, y_1) and (x_2, y_2).

2. **Apply the distance formula**:
   - Calculate the difference between the x-coordinates $(x_2 - x_1)$.
   - Calculate the difference between the y-coordinates $(y_2 - y_1)$.
   - Square both differences.
   - Add the squared differences.
   - Take the square root of the sum to find the distance.

Example:

Find the distance between the points $(2, 1)$ and $(-1, 5)$.

Solution:

1. Identify the coordinates:
   $(x_1, y_1) = (2, 1)$
   $(x_2, y_2) = (-1, 5)$

2. Apply the distance formula:
   $d = \sqrt{(x_2 - x_1)^2 + (y_2 - y_1)^2}$

3. Substitute the coordinates into the formula:
   $d = \sqrt{(-1 - 2)^2 + (5 - 1)^2}$

4. Calculate the differences:
   $d = \sqrt{3^2 + 4^2} = \sqrt{9 + 16} = \sqrt{25} = 5$

So, the distance between the points $(2, 1)$ and $(-1, 5)$ is 5 units.

www.mathnotion.com

Word Problems

Steps to Solve Word Problems About Linear Functions

1. **Read the Problem Carefully:**

    - Understand what the problem is asking.
    - Identify the key information and quantities involved.

2. **Define the Variables:**

    - Assign variables to the quantities that change.
    - Typically, you'll use x for the independent variable and y for the dependent variable.

3. **Set Up the Equation:**

    - Determine the relationship between the variables.
    - Use the given information to form a linear equation in the form $y = mx + b$.

4. **Solve the Equation:**

    - Substitute the known values into the equation to find the unknowns.
    - Perform the necessary algebraic steps to solve the variables.

5. **Interpret the Solution:**

    - Ensure your solution makes sense in the context of the problem.
    - Answer the question posed by the problem.

Example

A taxi company charges a base fare of $3 plus $2 per mile driven. Write a linear equation that represents the total fare (y) in terms of the number of miles driven (x). Then, find the fare for a 10-mile trip.

Solution:

1. Read the problem carefully: Base fare: $3, Cost per mile: $2, Total fare: y, Miles driven: x

2. Define the variables: Let x be the number of miles driven and let y be the total fare.

3. Set up the equation: The total fare is the base fare plus the cost per mile times the number of miles: $y = 2x + 3$

4. Solve the equation for a Specific Value: For a 10-mile trip ($x = 10$): $y = 2(10) + 3 = 23$

5. Interpret the solution: The fare for a 10-mile trip is $23.

Worksheets

🕮 Slope from Graph
Find the slope of each graph.

1)
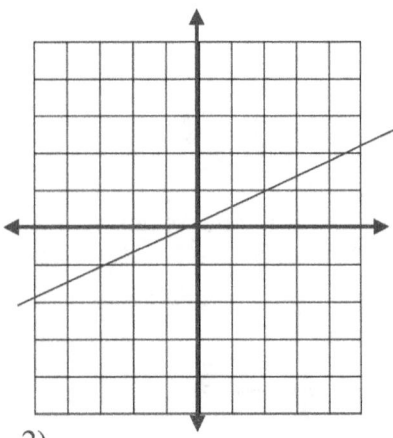

2)
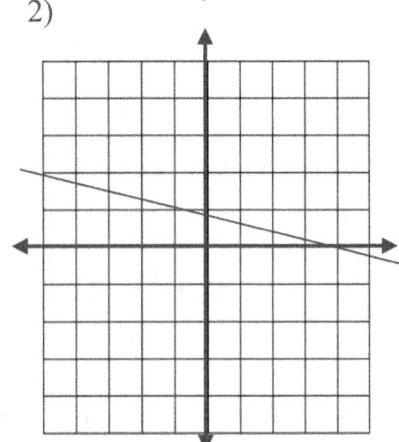

3)
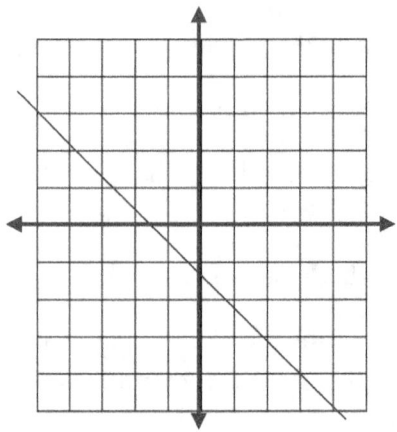

4)
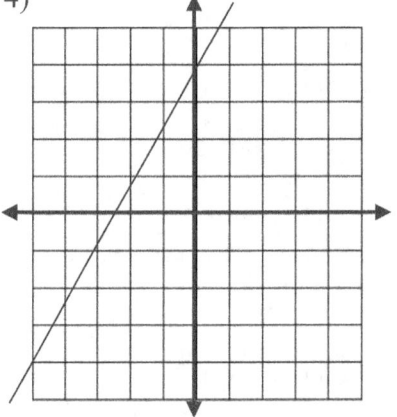

🕮 Slope from Two Points
Find the slope of two points.
1) $(1,3)$ and $(-1,6)$
2) $(4,2)$ and $(5,2)$
3) $(-2,3.5)$ and $(-1,2.5)$
4) $(1.5,4)$ and $(2,3.2)$
5) $\left(\frac{2}{3},\frac{6}{5}\right)$ and $\left(\frac{-4}{3},\frac{3}{10}\right)$

🕮 Identifying Slope and Y-Intercept
Identify slope and y-intercept.
1) $y = -2x + 5$
2) $3y - x = -1$
3) $\frac{4}{5}y + 10x + 1 = 0$
4) $y = -\frac{x}{3}$
5) $2y + 2.5x + 1 = 0$
6) $6y - 3 = 0$
7) $3 + \frac{2y}{5} = 15x$
8) $-5x + \frac{y}{3} + \frac{1}{2} = 0$
9) $-y - \frac{4x}{5} = 0$
10) $2 + 1.2x - 0.6y = 0$

www.mathnotion.com

Rate of Change

What is the average rate of change?
1) $f(x) = 4x - 2$ from $x = -2.5$ to $x = 3$
2) $f(x) = x^3 + 2x - 1$ from $x = 0$ to $x = 2$
3) $f(x) = 2x^2 - 4x$ from $x = \frac{1}{2}$ to $x = 4$
4) $f(x) = \frac{x^3+1}{2}$ from $x = 1$ to $x = 5$
5) $f(x) = \frac{1}{4}x^2 - 2x + 3$ from $x = -2$ to $x = 6$

Graphing Lines Using Line Equation

Graph following line equations:
1) $3y = 6$
2) $2y - x = 3$
3) $3y + \frac{x}{5} = 0$
4) $-y + 4x + 3 = 0$

Writing Linear Equations

Write the equation of the corresponding line based on the given information
1) Write the equation of a line with a slope of 3 and a y-intercept of 5.
2) Write the equation of a line with a slope of $\frac{-2}{3}$ and passing through the y-axis at -4.
3) Write the equation of a line that passes through the origin with a slope of $\frac{1}{2}$.
4) Find the equation of a line that passes through the points $(2, -2)$ and $(0, 5)$.
5) Write the equation of a line that has a y-intercept of 7 and is parallel to the line $y = -x + 2$.

Write an Equation from Graph

Write the equation of the following lines:

1)
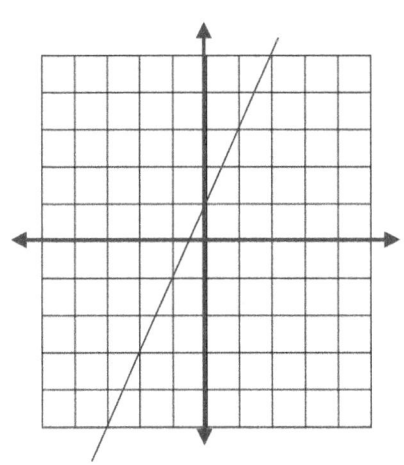

2)
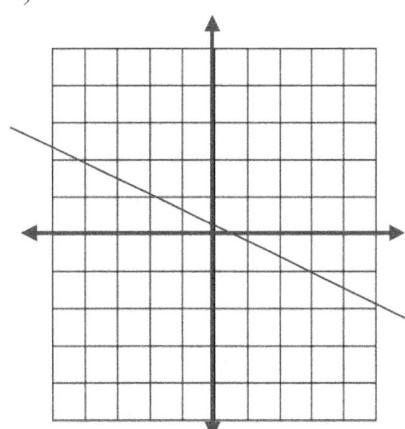

3) 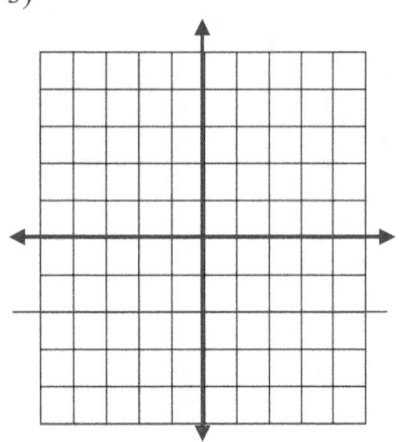   4)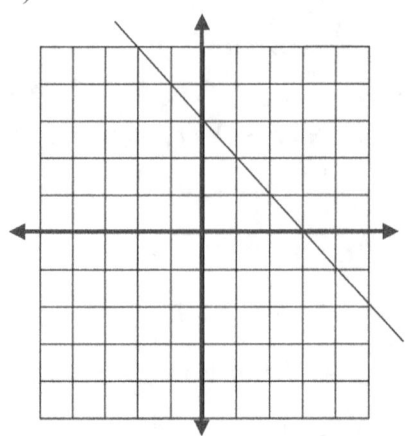

✎ Graphing Linear Inequalities

Graph the inequalities.
1) $y > 2x - 5$
2) $y \geq -x + 4$
3) $-y + 3 < x$
4) $5y + 10x \leq 20$

✎ System of Equations

Solve the system of equations.

1) $\begin{cases} y - x = 3 \\ 2y + x = -1 \end{cases}$

2) $\begin{cases} 5y - 3x = 1 \\ 2y + x = -2 \end{cases}$

3) $\begin{cases} -3y + 3x = 6 \\ 2y + 3x = -4 \end{cases}$

4) $\begin{cases} 2y + 4x = 7 \\ -y - x = 5 \end{cases}$

5) $\begin{cases} 6y + x = -8 \\ 5y - 4x = 10 \end{cases}$

✎ Point Coordinate and Finding Midpoint

Find the midpoint.
1) $(2, -5)$ and $(0, 3)$
2) $(4, 3)$ and $(7, 1)$
3) $(-1, 0)$ and $(2.6, 1.2)$
4) $(\frac{4}{5}, 3)$ and $(-\frac{1}{3}, -2)$
5) $\left(-2\frac{1}{5}, 1.8\right)$ and $(2.4, 3\frac{1}{10})$

✎ Finding Distance of Two Points

Find the distance between two points:
1) $(1, 4)$ and $(1, -8)$
2) $(2, -6)$ and $(9, 3)$
3) $(3.5, 1)$ and $(2.5, 1)$
4) $(\frac{1}{5}, -\frac{1}{2})$ and $(\frac{7}{10}, 2.5)$
5) $\left(6.1, -\frac{1}{4}\right)$ and $(1.1, \frac{9}{25})$

www.mathnotion.com

🏴 Word Problems

Do the word problems by writing linear equations:
1) The cost of renting a bike is $10 plus $2 for each hour. Write an equation to represent the total cost (y) for x hours of rent.
2) A phone company offers a plan with a monthly fee of $20 and charges $0.05 per text message. Write an equation to represent the total cost (y) for x text messages in a month and find the cost for 200 text messages.
3) A movie theater sells adult tickets for $12 and child tickets for $8. If the theater sells a total of 50 tickets and earns $500, how many adult tickets and child tickets were sold?
4) A farmer has chickens and cows. If there are a total of 20 animals and 56 legs, how many chickens and cows does the farmer have?
5) A company produces two types of products, A and B. Product A costs $5 to make, and Product B costs $7 to make. If the company produces 100 products and spends $620, how many of each product were made?
6) The coordinates of a square are $A: (-1, -1), B: (-1, 3), C: (3, 3)$ and $D: (3, -1)$. Find the length of the diagonal of the square?
7) The coordinate of a rectangle is $A: (2, -4), B: (2, 2), C: (6, -4)$ and $D: (6, 2)$ respectively. Find the distance between the midpoints of segments AB and BC.
8) The coordinates of a triangle are $A: (-1, -2), B: (-1, 4)$ and $C: (6, -2)$. Find the perimeter and area of triangle.
9) John is twice as old as his brother Alex. Five years ago, John was three times as old as Alex. How old are John and Alex now?
10) A rectangle has a perimeter of 40 meters and the length of rectangle is one unit more than twice its width. What is the length and width of the rectangle?

Answer of worksheets

Slope from Graph
1) $\frac{2}{5}$
2) $-\frac{2}{9}$
3) $\frac{-8}{9}$
4) $\frac{3}{2}$

Slope from Two Points
1) $\frac{-3}{2}$
2) 0
3) -1
4) -1.6
5) $\frac{9}{20}$

Identifying Slope and Y-Intercept from a Linear Equation
1) $m: -2$ and $b: 5$
2) $m: \frac{1}{3}$ and $b: \frac{-1}{3}$
3) $m: \frac{-25}{2}$ and $b: \frac{-5}{4}$
4) $m: \frac{-1}{3}$ and $b: 0$
5) $m: -1.25$ and $b: \frac{-1}{2}$
6) $m: 0$ and $b: \frac{1}{2}$
7) $m: \frac{75}{2}$ and $b: \frac{-15}{2}$
8) $m: 15$ and $b: \frac{-3}{2}$
9) $m: \frac{-4}{5}$ and $b: 0$
10) $m: 2$ and $b: \frac{10}{3}$

Graphing Lines Using Line Equation

1)
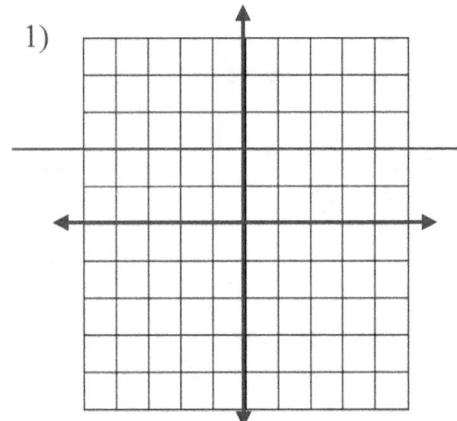

2)
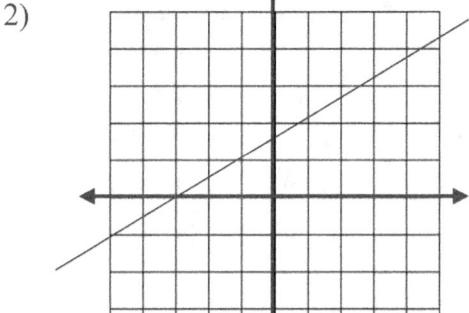

3)
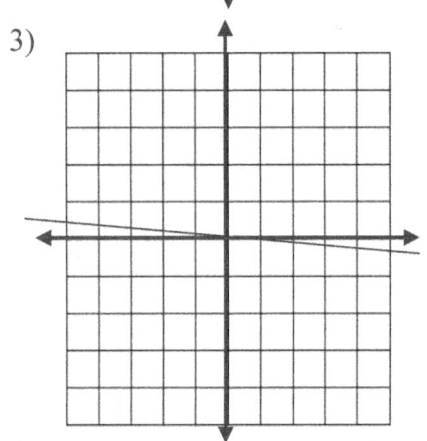

4)
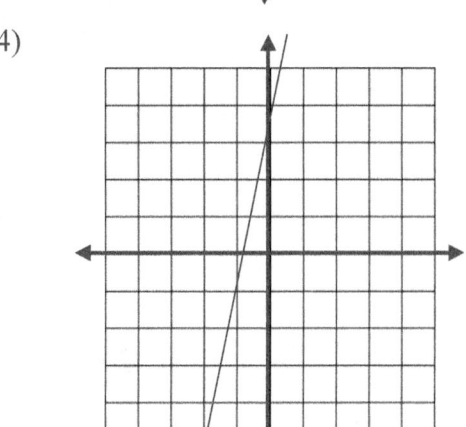

Rate of Change
1) 4
2) 6
3) 5
4) 15.5
5) −1

Writing Linear Equations
1) $y = 3x + 5$
2) $y = -\frac{2}{3}x - 4$
3) $y = \frac{1}{2}x$
4) $y = -\frac{7}{2}x + 5$
5) $y = -x + 7$

Write an Equation from Graph
1) $y = 2x + 1$
2) $y = -\frac{3}{7}x + \frac{2}{7}$
3) $y = -2$
4) $y = -x + 3$

Graphing Linear Inequalities

1)
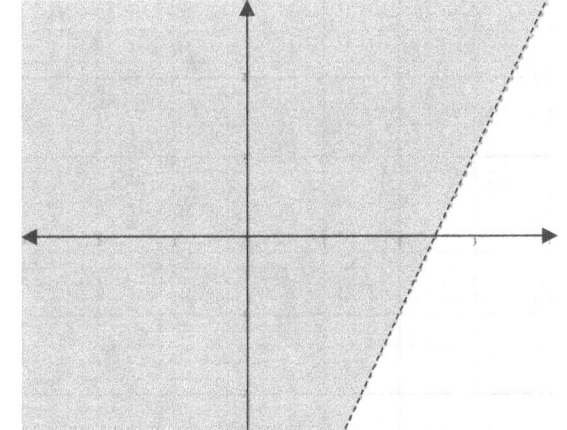

2)
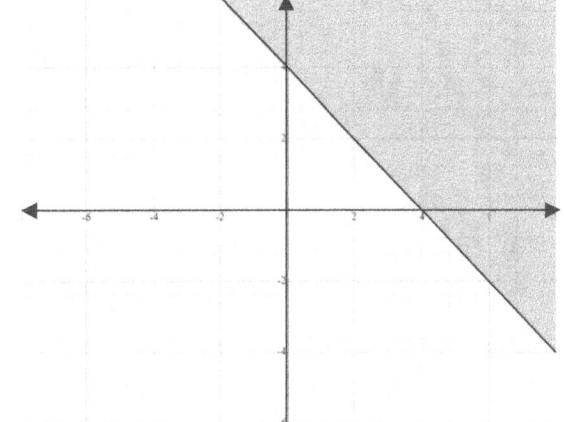

3)
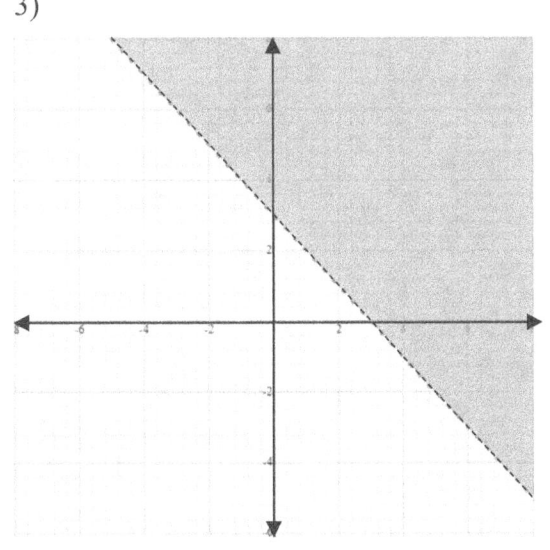

4)
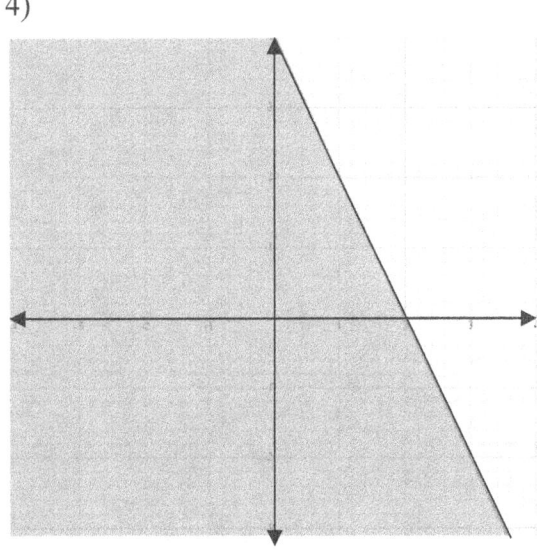

System of Equations

1) $\begin{cases} y = \frac{2}{3} \\ x = \frac{-7}{3} \end{cases}$

2) $\begin{cases} y = \frac{-5}{11} \\ x = \frac{-12}{11} \end{cases}$

3) $\begin{cases} y = -2 \\ x = 0 \end{cases}$

4) $\begin{cases} x = \frac{17}{2} \\ y = \frac{-27}{2} \end{cases}$

5) $\begin{cases} y = \frac{-22}{29} \\ x = \frac{-100}{29} \end{cases}$

Point Coordinate and Finding Midpoint

1) $(1, -1)$
2) $(\frac{11}{2}, 2)$
3) $(0.8, 0.6)$
4) $(\frac{7}{30}, \frac{1}{2})$
5) $(\frac{1}{10}, 2.45)$

Finding Distance of Two Points

1) $d = 12$
2) $d = \sqrt{130}$
3) $d = 1$
4) $d = \sqrt{9.25}$
5) $d = \sqrt{25.49}$

Word Problems

1) $y = 10 + 2x$
2) $y = 20 + 0.05x$ and total cost for 200 messages: $y = 30$
3) 25 adult tickets and 25 child tickets
4) 12 chickens and 8 cows
5) 40 products of type A and 60 products of type B
6) $4\sqrt{2}$ units
7) 2 units
8) Perimeter: $13 + \sqrt{85}$ units and Area: 21 square units
9) John: 20 years old and Alex: 10 years old
10) The width: $\frac{19}{3}$ and the length: $\frac{41}{3}$

Chapter 10: Transformations

Topics that you'll learn in this chapter:

- ✓ Translations
- ✓ Reflections
- ✓ Rotations
- ✓ Dilations
- ✓ Scale Drawings
- ✓ Worksheets
- ✓ Answer of Worksheets

Translations

In geometry, a translation is a type of transformation that slides each point of a shape or figure the same distance in the same direction. It's like picking up the figure and moving it without rotating, resizing, or flipping it. Translations maintain the shape's orientation and size, meaning the figure and its translated image are congruent.

Key Points:

1. **Direction and Distance**: Translation involves moving a figure along a straight path. The direction and distance of the translation are specified by a vector.
2. **Vector Notation**: A translation vector, often denoted as $\vec{v} = (a, b)$, specifies the movement. The components a and b represent the horizontal and vertical shifts, respectively.
3. **Coordinate Changes**: If a point (x, y) is translated by its new position (x', y') is given by:
$$x' = x + a$$
$$y' = y + b$$

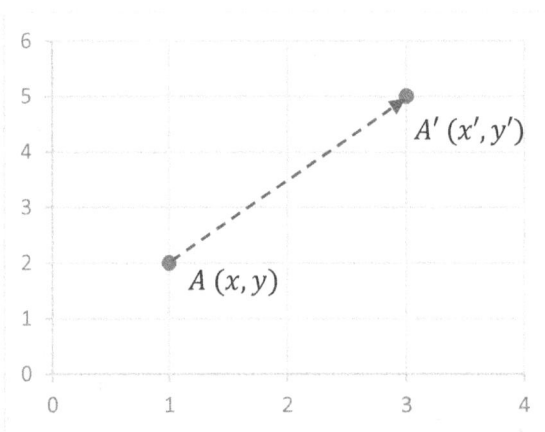

Example:

Consider a triangle with vertices at $A(1,2)$, $B(3,4)$, and $C(5,2)$. Let's translate the triangle by the vector $(2,3)$.

Solution:

1. Original points: $A(1,2)$, $B(3,4)$ and $C(5,2)$
2. Translation vector: $(2,3)$
3. New points after translation:
   - $A'(1+2, 2+3) = (3,5)$
   - $B'(3+2, 4+3) = (5,7)$
   - $C'(5+2, 2+3) = (7,5)$

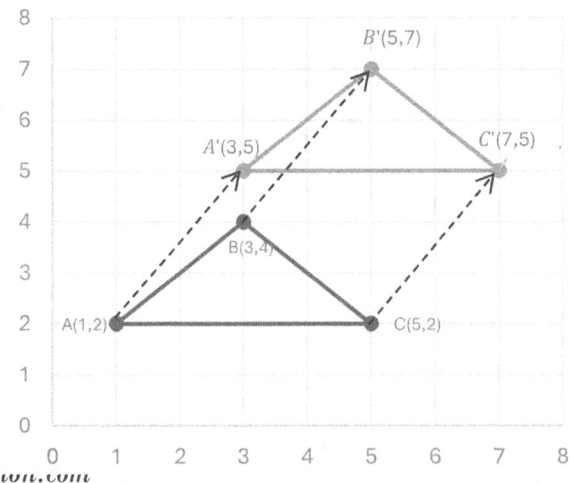

Reflections

In geometry, a reflection is a type of transformation that flips a figure over a line, creating a mirror image. The line of reflection acts as the "mirror" and each point on the figure and its reflected image are equidistant from this line. Reflections preserve the size and shape of the figure, making the original figure and its reflected image congruent.

Key Points:

1. **Line of Reflection**: The line over which the figure is reflected. Common lines of reflection include the $x-axis$, $y-axis$, and the line $y = x$.

2. **Coordinate Changes**: Depending on the line of reflection, the coordinates of the points change as follows:

   - **Reflection over the x-axis**: If a point (x, y) is reflected over the x-axis, its new coordinates (x', y') are: ($figure$ 1)
   $$x' = x \text{ and } y' = -y$$

   - **Reflection over the y-axis**: If a point (x, y) is reflected over the y-axis, its new coordinates (x', y') are: ($figure$ 2)
   $$x' = -x \text{ and } y' = y$$

   - **Reflection over the line $y = x$**: If a point (x, y) is reflected over the line $y = x$, its new coordinates (x', y') are: ($figure$ 3)
   $$x' = y \text{ and } y' = x$$

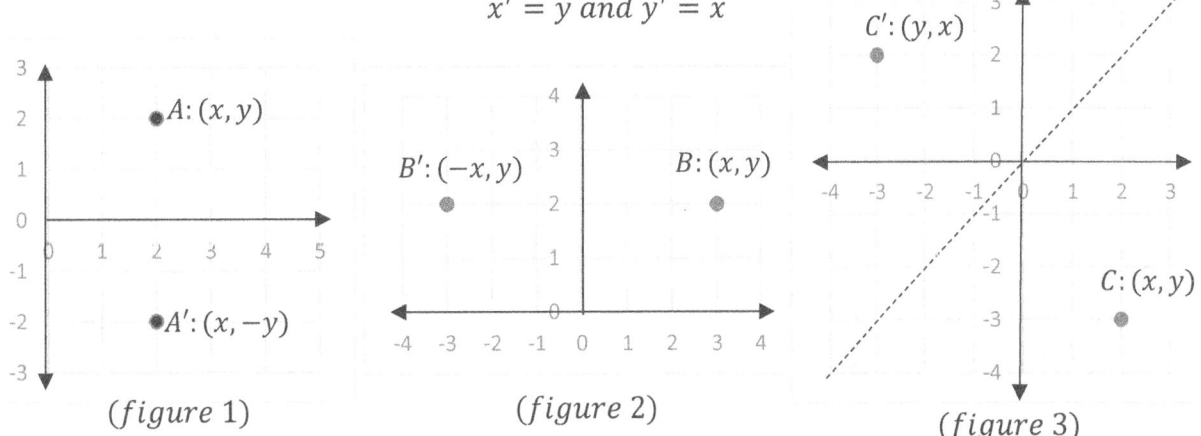

($figure$ 1) ($figure$ 2) ($figure$ 3)

Example:

Reflect point $A: (4, -3)$ over:

  a) Over the x-axis
  b) Over the y-axis
  c) Over the line $y = x$

Solution:
  a) The reflected point will be $A': (4, 3)$
  b) The reflected point will be $A': (-4, -3)$
  c) The reflected point will be $A': (-3, 4)$

Rotations

In geometry, a rotation is a type of transformation that turns a figure around a fixed point, called the center of rotation. This point remains stationary while the rest of the figure moves. The rotation is defined by three main elements: the center of rotation, the angle of rotation, and the direction of rotation (clockwise or counterclockwise).

Key Points:

1. **Center of Rotation**: The fixed point around which the figure rotates. Common centers of rotation include the origin (0,0) or any other specified point.

2. **Angle of Rotation**: The measure of the degree to which the figure is rotated around the center. It is typically measured in degrees (°), such as 90°, 180°, or 270°.

3. **Direction of Rotation**: The direction in which the figure is rotated. It can be either clockwise or counterclockwise.

Rotation Rules: The coordinates of a point (x, y) after a rotation around the origin $(0,0)$ are transformed as follows:

- **90° Counterclockwise Rotation**: $(x, y) \to (-y, x)$
- **180° Rotation** (same in both directions): $(x, y) \to (-x, -y)$
- **270° Counterclockwise Rotation** (or 90° Clockwise): $(x, y) \to (y, -x)$

Example:

Take a triangle with points (1,1), (2,1), and (1,2). Rotate it 90 degrees counterclockwise around the origin.
  Solution:
  - (1,1) becomes (−1,1)
  - (2,1) becomes (−1,2)
  - (1,2) becomes (−2,1)

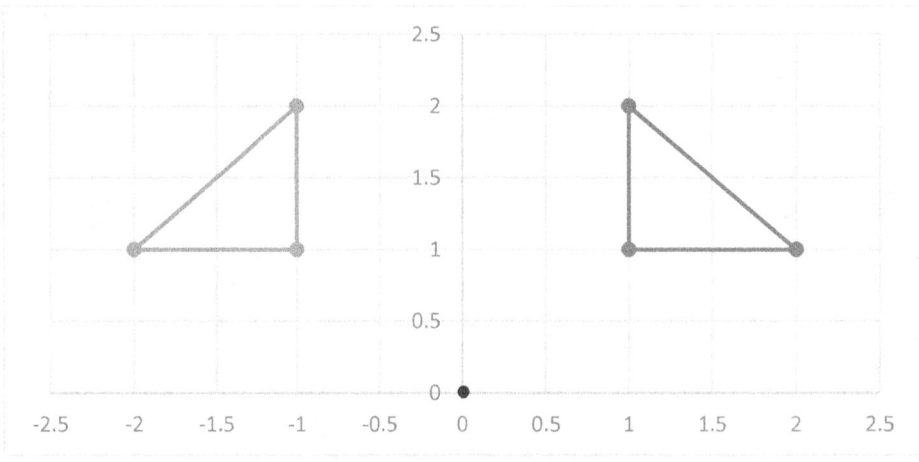

Dilations

In geometry, a dilation is a type of transformation that changes the size of a figure, but not its shape. The figure is either enlarged or reduced by a scale factor relative to a fixed point called the center of dilation. A dilation transformation preserves the angles of the figure, meaning the original figure and its dilated image are similar.

Key Points:

1. **Center of Dilation**: The fixed point around which the figure is dilated. This point remains unchanged during the transformation.

2. **Scale Factor**: A positive number that determines how much the figure is enlarged or reduced.

   - If the scale factor is greater than 1, the figure is enlarged.
   - If the scale factor is between 0 and 1, the figure is reduced.
   - If the scale factor is exactly 1, the figure remains the same size.

3. **Coordinate Changes**: If the center of dilation is the origin (0,0) and the scale factor is k, the coordinates of a point (x, y) after dilation are transformed as follows:

$$(x', y') = (kx, ky)$$

Example:

Take a triangle with points (1,1), (1,2) and (2,1), Dilate it with a scale factor of 3 from the origin.

Solution:

- (1,1) becomes (3,3)
- (1,2) becomes (3,6)
- (2,1) becomes (6,3)

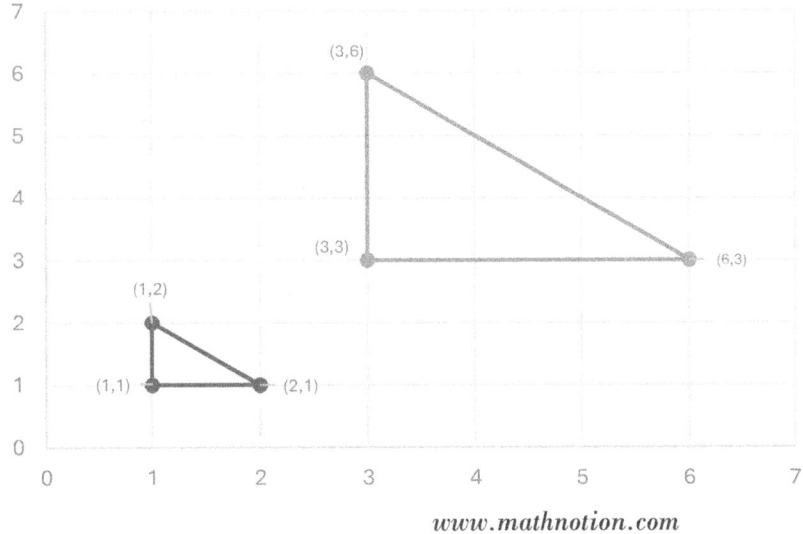

www.mathnotion.com

Scale Drawings

Scale drawings are representations of objects or structures where the dimensions are proportional to the actual object but reduced or enlarged by a certain factor called the scale. They are essential for accurately depicting large objects (like buildings or maps) on a smaller surface, or vice versa.

Key Concepts:

1. **Scale Factor**: The ratio of the dimensions in the drawing to the actual dimensions. It can be expressed as a fraction, ratio, or with a colon (e.g., $1:50$).
   - If the scale factor is less than 1 (e.g., $1:50$), the drawing is smaller than the actual object.
   - If the scale factor is greater than 1 (e.g., $50:1$), the drawing is larger than the actual object.

2. **Proportionality**: All dimensions in the scale drawing are proportional to the actual dimensions. This means that the ratios of corresponding lengths in the drawing and the actual object are equal.

3. **Applications**: Scale drawings are widely used in various fields such as architecture, engineering, and cartography (map making).

Example:

A city park is represented on a map with a scale of $1:500$. On the map, the length of the park is 10 cm and the width are 6 cm. Find the actual length and width of the park in meters.

Solution:

1. Determine the scale factor: The scale is $1:500$, which means 1 cm on the map represents 500 cm in real life.
2. Convert the length and width:
   - Length on the map: 10 cm
   - Width on the map: 6 cm
   - Scale factor: 500
3. Calculate actual length:
   - $Actual\ length = Length\ on\ the\ map \times Scale\ factor = 10\ cm \times 500 = 5{,}000\ cm$
   - Convert to meters: (Since $100\ cm = 1\ meter$): $5{,}000\ cm \div 100 = 50\ meters$
4. Calculate actual width:
   - $Actual\ width = width\ on\ the\ map \times Scale\ factor = 6\ cm \times 500 = 3{,}000\ cm$
   - Convert to meters: $3{,}000\ cm \div 100 = 30\ meters$

Worksheets

🔖 Translations

Find the translations.

1) Translate the triangle with vertices $A(1,2), B(4,3)$, and $C(3,7)$ by the vector $\vec{v} = (5,-2)$. What are the new coordinates of the vertices?
2) A rectangle has vertices at $(2,1), (2,4), (5,4)$ and $(5,1)$. Translate the rectangle by the vector $\vec{v} = (-3,6)$. What are the new coordinates of the vertices?
3) If the vertices of a parallelogram, after translation by $\vec{v} = (-1,3)$ match these vertices: $A(1,1), B(6,1), C(3,3)$ and $D(8,3)$, find the original vertices of the parallelogram.
4) If the vertices of a kite after translation by $\vec{v} = (-2,3)$ match these vertices: $A(0,0), B(3,4), C(6,0)$ and $D(3,-4)$, find the original vertices of the kite?
5) If point $A(2,5)$ is translated three times by three vectors: $\vec{a} = (1,2), \vec{b} = (-2,4), \vec{c} = (3,5)$, find the final translated point.

🔖 Reflections

Find the reflections.

1) Reflect the point $P(2,3)$ over the x-axis. What are the coordinates of the reflected point?
2) Reflect the point $Q(-4,-5)$ over the y-axis. What are the coordinates of the reflected point?
3) Reflect the point $R(3,7)$ over the line $y = x$. What are the coordinates of the reflected point?
4) A triangle has vertices $A(1,1), B(3,4)$, and $C(5,2)$. Reflect the triangle over the line $y = 2$. What are the coordinates of the reflected vertices?
5) Reflect the point $T(6,-3)$ over the line $y = -x$. What are the coordinates of the reflected point?

🔖 Rotations

Do the rotations.

1) Rotate the point $P(2,3)$ 90° counterclockwise around the origin. What are the coordinates of the rotated point?
2) Rotate the point $Q(-4,-5)$ 180° around the origin. What are the coordinates of the rotated point?
3) Rotate the point $R(3,7)$ 270° counterclockwise around the origin. What are the coordinates of the rotated point?
4) Rotate the pentagon with vertices $A(1,2), B(3,5), C(5,4), D(4,1)$, and $E(2,0)$ 90° clockwise around the origin. What are the coordinates of the rotated vertices?
5) Rotate the quadrilateral with vertices $P(2,2), Q(4,2), R(4,5)$, and $S(2,5)$ 180° around the origin. What are the new coordinates of the vertices?

www.mathnotion.com

Dilations

Find the dilations.

1) A triangle has vertices $A(1, 2), B(3, 4)$, and $C(5, 6)$. If the triangle is dilated from the origin with a scale factor of 0.5, what are the new coordinates of the vertices?
2) A pentagon has vertices $A(0, 0), B(2, 3), C(4, 3), D(5, 1)$, and $E(3, -2)$. If the pentagon is dilated from the origin with a scale factor of 2, what are the new coordinates of the vertices?
3) A parallelogram has vertices $A(2, 1), B(5, 3), C(7, 2)$, and $D(4, 0)$. Dilate the parallelogram from the origin with a scale factor of 4. What are the new coordinates of the vertices?
4) A triangle has vertices $A(-3, 4), B(0, -2)$, and $C(5, 2)$. If the triangle is dilated from the origin with a scale factor of 1.25, what are the new coordinates of the vertices?
5) A square has one vertex at $(1, 1)$ and the center at the origin. Dilate the octagon from the origin with a scale factor of 0.5. What are the new coordinates of all the vertices?

Scale Drawings

Do the scale drawings.

1) On a map with a scale of $1: 100$, a park is represented by a rectangle measuring 4 cm by 6 cm. What are the actual dimensions of the park in meters?
2) A blueprint of a house has a scale of $1: 50$. If a room is 3 cm wide on the blueprint, what is the actual width of the room in meters?
3) On a map with a scale of $1: 200$, a river is represented as a line segment 8 cm long. What is the actual length of the river in kilometers?
4) A blueprint of a new office building shows a room that measures 5 cm by 8 cm on the blueprint. The actual dimensions of the room are 10 meters by 16 meters. What is the scale of the blueprint?
5) A model of a famous monument is created with a scale factor. The actual height of the monument is 120 meters, while the height of the model is 15 cm. Determine the scale of the model.

Answer of Worksheets

Translations
1) $A' = (6, 0), B' = (9, 1),$ and $= C'(8, 5)$.
2) $(2, 1) \rightarrow (-1, 7)$
   $(2, 4) \rightarrow (-1, 10)$
   $(5, 4) \rightarrow (2, 10)$
   $(5, 1) \rightarrow (2, 7)$
3) $A' = (2, -2), B' = (7, -2), C' = (4, 0)$ and $D' = (9, 0)$
4) $A' = (2, -3), B' = (5, 1), C' = (8, -3)$ and $D' = (5, -7)$
5) $A' = (4, 16)$

Reflections
1) $P' = (2, -3)$
2) $Q' = (4, -5)$
3) $R' = (7, 3)$
4) $A' = (1, 3), B' = (3, 0),$ and $C' = (5, 2)$
5) $T' = (3, -6)$

Rotations
1) $P' = (-3, 2)$
2) $Q' = (4, 5)$
3) $R' = (7, -3)$
4) $A' = (2, -1), B' = (5, -3), C' = (4, -5), D' = (1, -4),$ and $E' = (0, -2)$
5) $P' = (-2, -2), Q' = (-4, -2), R' = (-4, -5),$ and $S' = (-2, -5)$

Dilations
1) $A' = (0.5, 1), B' = (1.5, 2),$ and $C' = (2.5, 3)$
2) $A' = (0, 0), B' = (4, 6), C' = (8, 6), D' = (10, 2),$ and $E' = (6, -4)$
3) $A' = (8, 4), B' = (20, 12), C' = (28, 8),$ and $D' = (16, 0)$
4) $A' = (-3.75, 5), B' = (0, -2.5),$ and $C' = (6.25, 2.5)$
5) $(0.5, 0.5), (-0.5, 0.5), (-0.5, -0.5)$ and $(0.5, -0.5)$

Scale Drawings
1) The actual dimensions of the park are 4 meters by 6 meters
2) 1.5 meters
3) 0.016 kilometers
4) 1 : 200
5) 1 : 800

Chapter 11: Sequences

Topics that you'll learn in this chapter:
- ✓ Identify Arithmetic Sequences
- ✓ Identify Geometric Sequences
- ✓ Evaluate Variable Expressions for Sequences
- ✓ Word Problems
- ✓ Worksheets
- ✓ Answer of Worksheets

Identify Arithmetic Sequences

An arithmetic sequence (or arithmetic progression) is a sequence of numbers in which the difference between consecutive terms is constant. This difference is called the common difference.

Key Characteristics:

1. **Common Difference (d)**: The difference between any two successive terms in the sequence.

2. **First Term (a_1)**: The initial term of the sequence.

General Formula:

The n-th term of an arithmetic sequence can be found using the formula:
$$a_n = a_1 + (n-1)d$$

Where:

- a_n is the n-th term.
- a_1 is the first term.
- d is the common difference.
- n is the position of the term in the sequence.

Identifying Arithmetic Sequences: To identify whether a sequence is arithmetic, you can:

1. Calculate the difference between successive terms.

2. Check if this difference is constant throughout the sequence.

Examples:

1) Identify if the sequence 3, 7, 11, 15, ... is arithmetic or not.
   Solution:
   1. Calculate the differences between successive terms:
      - $7 - 3 = 4$
      - $11 - 7 = 4$
      - $15 - 11 = 4$
   2. Since the difference is constant (4), it is an arithmetic sequence.

2) If the sequence 8, 13, 18, 23, ... is arithmetic, identify the a_1, d and a_n.
   Solution:
   - The first term (a_1) is 8
   - The common difference (d) is 5 (since $13 - 8 = 5$)
   - The general formula is:
     $a_n = a_1 + (n-1)d = 8 + (n-1)5 = 8 + 5n - 5 = 3 + 5n$
     So, the general formula is $a_n = 3 + 5n$

Identify Geometric Sequences

A geometric sequence (or geometric progression) is a sequence of numbers where each term after the first is obtained by multiplying the previous term by a constant called the common ratio.

Key Characteristics:

1. **Common Ratio (r):** The ratio between any two successive terms in the sequence.
2. **First Term (a_1):** The initial term of the sequence.

General Formula:

The *n-th* term of a geometric sequence can be found using the formula:

$$a_n = a_1 \times r^{n-1}$$

Where:

- a_n is the *n-th* term.
- a_1 is the first term.
- r is the common ratio.
- n is the position of the term in the sequence.

Identifying Geometric Sequences: To identify whether a sequence is geometric, you can:

1. Calculate the ratio between successive terms.
2. Check if this ratio is constant throughout the sequence.

Examples:

1) Identify if the sequence $3, 9, 27, 8, 1 \ldots$ is geometric or not.
   Solution:
   1. Calculate the ratio between successive terms:
      - $9 \div 3 = 3$
      - $27 \div 9 = 3$
      - $81 \div 27 = 3$
   2. Since the ratio is constant (3), it is a geometric sequence.

2) If the sequence $1, 4, 16, 64, \ldots$ is geometric, identify the a_1, r and a_n.
   Solution:
   - The first term (a_1) is 1
   - The common ratio (r) is 4 (since $4 \div 1 = 4$)
   - The general formula is:
     $a_n = a_1 \times r^{n-1} = 1 \times 4^{n-1} = 4^{n-1}$

So, the general formula is $a_n = 4^{n-1}$

Evaluate Variable Expressions

Evaluating variable expressions for sequences involves substituting values into the expression and simplifying it according to the rules of algebra.

Steps to Evaluate Variable Expressions:

1. **Identify the sequence type** (arithmetic or geometric).
2. **Determine the general formula** for the sequence.
3. **Substitute the given values** into the formula.
4. **Simplify the expression** to find the desired term.

Examples:

1) Find the 20-th term of sequence: $2, 7, 12, 17, 22, ...$

   Solution:

   1. Identify the sequence type: Since the difference between successive terms is constant ($5: 7 - 2 = 5, 12 - 7 = 5, ...$), so the sequence is arithmetic.
   2. Determine the general formula for the sequence and substitute the values into the formula: The first term (a_1) is 2, the common difference (d) is 5 and $a_n = 2 + (n - 1)5 = 5n - 3$
   3. Simplify the expression to find the desired term:

      $a_n = 5n - 3 \rightarrow a_{20} = 5 \times 20 - 3 = 100 - 3 = 97$

      So, $a_{20} = 97$

2) Find the 10-th term of sequence: $6, 3, \frac{3}{2}, \frac{3}{4}, \frac{3}{8}, ...$

   Solution:

   1. Identify the sequence type: Since the ratio between successive terms is constant ($\frac{1}{2}: 3 \div 6 = \frac{1}{2}, \frac{3}{2} \div 3 = \frac{1}{2}, ...$), so the sequence is geometric.
   2. Determine the general formula for the sequence and substitute the values into the formula: The first term (a_1) is 6, the ratio (r) is $\frac{1}{2}$ and $a_n = 6 \times (\frac{1}{2})^{n-1}$
   3. Simplify the expression to find the desired term:

      $a_n = 6 \times (\frac{1}{2})^{n-1} \rightarrow a_{10} = 6 \times (\frac{1}{2})^{10-1} = 6 \times (\frac{1}{2})^9 = 6 \times \frac{1}{512} = \frac{6}{512} = \frac{3}{256}$

      So, $a_{10} = \frac{3}{256}$

Word Problems

Steps to Solve Word Problems About Sequences:

1. **Understand the Problem:**

    - Carefully read the problem and identify what is being asked.
    - Determine if the sequence is arithmetic, geometric, or another type.

2. **Identify Key Information:** Extract important details such as the first term, common difference (arithmetic sequence), common ratio (geometric sequence), or specific term positions.

3. **Determine the Formula:**

    - For arithmetic sequences: $a_n = a_1 + (n-1)d$
    - For geometric sequences: $a_n = a_1 \times r^{n-1}$

4. **Set Up the Equation:** Use the information from the problem to create an equation based on the appropriate formula.

5. **Solve the Equation:** Substitute the known values into the equation and solve for the unknown variable.

6. **Check Your Solution:** Verify that your solution makes sense in the context of the problem and that all parts of the question are addressed.

Example:

The theater has 20 rows of seats. Each row has 2 more seats than the previous row. The first row has 15 seats. How many seats are in the $15th$ row?

Solution:

1. Identify the sequence type: This is an arithmetic sequence because each row has a constant difference in the number of seats.
2. Extract key information:
    - First term (a_1): 15 seats
    - Common difference (d): 2 seats
    - Term to find ($n = 15$)
3. Use the arithmetic sequence formula: $a_n = a_1 + (n-1)d$
4. Set up the equation: $a_{15} = 15 + (15 - 1) \times 2$
5. Solve the equation: $a_{15} = 15 + (15 - 1) \times 2 = 15 + 14 \times 2 = 15 + 28 = 43$
6. Check the solution: The 15th row has 43 seats, which makes sense given the increasing pattern.

Worksheets

📌 Identify Arithmetic sequences

Determine whether the sequences are arithmetic or not.

1) $5, 8, 11, 14, ...$
2) $3, 4, 6, 9, ...$
3) $-12, -12, -12, -12, ...$
4) $\frac{1}{2}, \frac{1}{6}, \frac{1}{10}, \frac{1}{14}, ...$
5) $9, 8.75, 8.5, 8.25, ...$

Identify first term (a_1), n-th term (a_n) and common difference (d).

6) $1, 5, 9, 13, ...$
7) $5, 7, 9, 11, ...$
8) $\frac{1}{10}, \frac{3}{10}, \frac{5}{10}, \frac{7}{10}, ...$
9) $1.25, 1.75, 2.25, 2.75, ...$
10) $11, 6, 1, -4, ...$

📌 Identify Geometric sequences

Determine whether the sequences are geometric or not.

1) $88, 88, 88, 88, ...$
2) $7, 14, 21, 28, ...$
3) $2, 6, 18, 54, ...$
4) $8, 4, 2, 1, ...$
5) $4.25, 8.25, 16.25, 32.25, ...$

Identify first term (a_1), n-th term (a_n) and the ratio (r).

6) $1, 9, 81, 729, ...$
7) $5, -10, 20, -40, ...$
8) $\frac{3}{5}, \frac{3}{10}, \frac{3}{20}, \frac{3}{40}, ...$
9) $1, 0.1, 0.01, 0.001, ...$
10) $15, 9, \frac{27}{5}, \frac{81}{5}, ...$

📌 Evaluate variable expressions

Find the requested term.

1) $a_n = 1 - 4n$ (find the a_{15})
2) $a_n = -7n + 8$ (find the a_{10})
3) $8, 10, 12, 14, ...$ (find the a_{20})
4) $-9, -6, -3, 0, ...$ (find the a_{100})
5) $24, 29, 33, 37, ...$ (find the a_{30})

Find the requested term in the following geometric sequences:

6) $a_n = 2 \times 5^{n-1}$ (find the a_6)

7) $a_n = \frac{2^{n-1}}{3}$ (find the a_{11})
8) $625, 125, 25, 5, \ldots$ (find the a_{10})
9) $-1, 1, -1, 1, \ldots$ (find the a_{1001})
10) $30000, 3000, 300, 30, \ldots$ (find the a_7)

Word Problems

Do the sequences word problems.

1) A sequence starts with -1 and each term increases by 5. What is the $12th$ term of the sequence?
2) A geometric sequence starts with 4 and has a common ratio of 0.5. What is the $5th$ term?
3) The population of a small town is 10,000 people. Every year, the population increases by 3%. How many people will live in the town after 3 years?
4) In an arithmetic sequence, the $5th$ term is 20 and the 10th term is 35. Find the first term and the common difference.
5) Alice invests $500 in a savings account. The account earns 4% interest annually, compounded once per year. How much money will she have in the account after 3 years?
6) A geometric sequence has the $3rd$ term 16 and the $6th$ term 128. What is the first term and the common ratio?
7) The sum of the first 10 terms of an arithmetic sequence is 140, and the first term is 5. Find the common difference.
8) A car rental company charges an initial fee of $50 and an additional $20 per day. Write the sequence that represents the total cost for the first 7 days of renting the car. What will the total cost be after 7 days?
9) A concert hall sold 200 tickets on the first day of sales. Each subsequent day, the number of tickets sold increases by 50. How many tickets will be sold by the $10th$ day?
10) A sequence follows the rule: $a_1 = 2, a_2 = 5$ and $a_n = a_{n-1} + a_{n-2}$ for $n \geq 3$ What is the $8th$ term in the sequence? (this sequence is neither arithmetic nor geometric sequence).

Answer of Worksheets

Identify Arithmetic sequences
1) Arithmetic.
2) Not arithmetic.
3) Arithmetic.
4) Not arithmetic.
5) Arithmetic.
6) $a_1 = 1, d = 4$ and $a_n = 4n - 3$
7) $a_1 = 5, d = 2$ and $a_n = 2n + 3$
8) $a_1 = \frac{1}{10}, d = \frac{1}{5}$ and $a_n = \frac{2n-1}{10}$
9) $a_1 = 1.25, d = 0.5$ and $a_n = 0.5n + 0.75$
10) $a_1 = 11, d = -5$ and $a_n = -5n + 16$

Identify Geometric sequences
1) Geometric.
2) Not geometric.
3) Geometric.
4) Geometric.
5) Not geometric.
6) $a_1 = 1, r = 9$ and $a_n = 9^{n-1}$
7) $a_1 = 5, r = -2$ and $a_n = 5 \times (-2)^{n-1}$
8) $a_1 = \frac{3}{5}, r = \frac{1}{2}$ and $a_n = \frac{3}{5} \times (\frac{1}{2})^{n-1}$
9) $a_1 = 1\ r = 0.1$ and $a_n = (0.1)^{n-1}$
10) $a_1 = 15, r = \frac{3}{5}$ and $a_n = 15 \times (\frac{3}{5})^{n-1}$

Evaluate variable expressions
1) $a_{15} = -59$
2) $a_{10} = -62$
3) $a_{20} = 46$
4) $a_{100} = 288$
5) $a_{30} = 169$
6) $a_6 = 6,250$
7) $a_{11} = \frac{1,024}{3}$
8) $a_{10} = \frac{1}{3,125}$
9) $a_{1001} = -1$
10) $a_7 = 0.03$

Word Problems
1) $a_{12} = 54$
2) $a_5 = 0.25$
3) Approximately 10,927 people.
4) $a_1 = 8\ and\ d = 3$.
5) Approximately $562.46.
6) $a_1 = 4\ and\ r = 2$.
7) $d = 2$.
8) $190.
9) 4,250 tickets.
10) $a_8 = 81$.

Chapter 12: Congruence and Similarity

Topics that you'll learn in this chapter:

- ✓ Similar Figures
- ✓ Ratio of Area and Volume in similar Figures
- ✓ Similarity Criteria for Triangles
- ✓ Similar Figures and Indirect Measurement
- ✓ Congruent Figures
- ✓ Congruence Criteria of Triangles
- ✓ Word Problems
- ✓ Worksheets
- ✓ Answer of Worksheets

Similar Figures

Similar figures are geometric shapes that have the same shape but different sizes. This means they have the same angles, and their corresponding sides are proportional. This means that one figure can be obtained from the other by a uniform scaling (enlarging or reducing), possibly with a translation, rotation, or reflection.

Key Properties of Similar Figures

1. **Corresponding Angles**: Corresponding angles of similar figures are equal.
2. **Proportional Sides**: The lengths of corresponding sides of similar figures are proportional.

Notation: If two figures are similar, we use the symbol ~ to denote this. For example, if triangle $\triangle ABC$ is similar to triangle $\triangle DEF$, we write:

$$\triangle ABC \sim \triangle DEF$$

Proportions in Similar Figures

If two figures are similar, the ratio of the lengths of corresponding sides is constant. For example, if the sides of triangle $\triangle ABC$ are a, b and c and the sides of triangle $\triangle DEF$ are d, e and f then:

$$\frac{a}{d} = \frac{b}{e} = \frac{c}{f} = k$$

where k is the **scale factor.**

Example:

Assume the two triangles below are similar, find the variable values and scale factor.

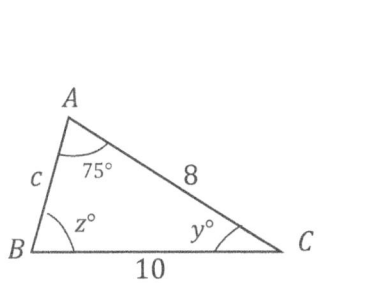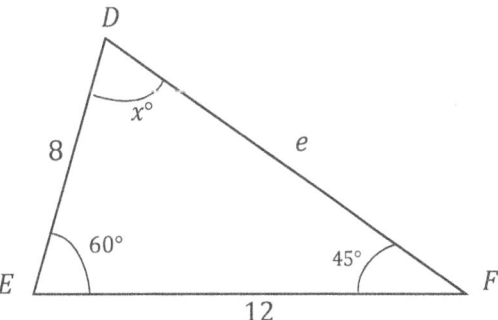

Solution:

- Since corresponding angles of similar figures are equal so: $x = 75°, y = 45°$ and $z = 60°$
- The ratio of corresponding sides is constant:

$$\frac{BC}{EF} = \frac{AC}{DF} = \frac{AB}{DE} = k \rightarrow \frac{10}{12} = \frac{8}{e} = \frac{c}{8} = k \rightarrow k \ (scale\ factor) = \frac{10}{12} = \frac{5}{6}$$

$$\frac{8}{e} = \frac{5}{6} \rightarrow 8 \times 6 = 5 \times e \rightarrow 48 = 5e \rightarrow e = \frac{48}{5} = 9.6$$

$$\frac{c}{8} = \frac{5}{6} \rightarrow 8 \times 5 = c \times 6 \rightarrow 40 = 6c \rightarrow c = \frac{40}{6} \approx 6.6$$

Ratio of Area and Volume in Similar Figures

Ratios of Areas:
- For two similar figures, the ratio of their areas is the square of the ratio of their corresponding linear dimensions (sides).
- If two similar figures have a linear ratio of k, then their area ratio is k^2
$$\frac{\text{Area of Figure 1}}{\text{Area of Figure 2}} = k^2$$

Ratios of Volumes:
- For two similar three-dimensional shapes, the ratio of their volumes is the cube of the ratio of their corresponding linear dimensions.
- If two similar shapes have a linear ratio of k, then their volume ratio is k^3.
$$\frac{\text{Volume of Figure 1}}{\text{Volume of Figure 2}} = k^3$$

Examples:

1) Two similar rectangles have corresponding side lengths in the ratio of $4:3$. If the area of the larger rectangle is $96\ cm^2$, what is the area of the smaller rectangle?
Solution:
 1. Identify the ratio of the side lengths: $4:3$ or $k = \frac{4}{3}$.
 2. Use the ratio of areas for similar figures, which is the square of the side ratio:
 $Area\ ratio = (\frac{4}{3})^2 = \frac{16}{9}$
 3. Let A be the area of the smaller rectangle. Since the area of the larger rectangle is $96\ cm^2$:
 $\frac{96}{A} = \frac{16}{9} \rightarrow 96 \times 9 = 16 \times A \rightarrow 864 = 16A \rightarrow A = \frac{864}{16} = 54$
 So, the area of the smaller rectangle is $54\ cm^2$.

2) Two similar spheres have radii in the ratio of $2:5$. If the volume of the smaller sphere is $32 cm^3$, what is the volume of the larger sphere?
Solution:
 1. Identify the ratio of the radii: $2:5$ or $k = \frac{2}{5}$.
 2. Use the ratio of volumes for similar figures, which is the cube of the side ratio:
 $Volume\ ratio = (\frac{2}{5})^3 = \frac{8}{125}$
 3. Let V be the volume of the larger sphere. Since the volume of the smaller sphere is $32\ cm^3$:
 $\frac{32}{V} = \frac{8}{125} \rightarrow 125 \times 32 = 8 \times V \rightarrow 4,000 = 8V \rightarrow V = \frac{4,000}{8} = 500\ cm^3$
 So, the volume of the larger sphere is $500\ cm^3$.

Similarity Criteria for Triangles

To determine if the two triangles are similar, we use specific criteria based on the relationships between their angles and sides. Here are the main criteria:

***AA (Angle − Angle)* Similarity**: If two angles of one triangle are equal to two angles of another triangle, the triangles are similar.

$$\angle A = \angle D \text{ and }, \angle B = \angle E \Rightarrow \triangle ABC \sim \triangle DEF$$

***SSS (Side − Side − Side)* Similarity**: If the corresponding side lengths of two triangles are proportional, then the triangles are similar.

$$\frac{AB}{DE} = \frac{BC}{EF} = \frac{CA}{FD} \Rightarrow \triangle ABC \sim \triangle DEF$$

This Criterion requires that all three pairs of corresponding sides are in the same ratio.

***SAS (Side − Angle − Side)* Similarity**: If two sides of one triangle are proportional to two sides of another triangle and the included angle between those sides is equal, the triangles are similar.

$$\frac{AB}{DE} = \frac{BC}{EF} \text{ and } \angle B = \angle E \Rightarrow \triangle ABC \sim \triangle DEF$$

Example:

Prove that $\triangle ABC \sim \triangle DEF$ given the following information in shapes:

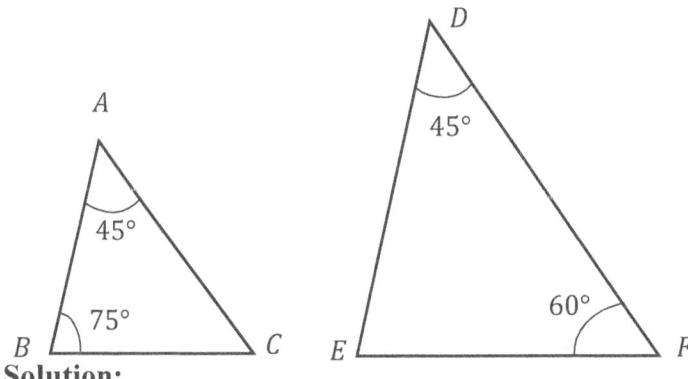

Solution:

- Given the information above:
  In $\triangle ABC$ we have: $\angle A = 45°, \angle B = 75°$ and $\angle C = 180° - (45° + 75°) = 60°$
  In $\triangle DEF$ we have: $\angle D = 45°, \angle F = 60°$ and $\angle E = 180° - (45° + 60°) = 75°$
- Apply the *AA* criterion: Given $\angle A = \angle D = 45°$ and $\angle C = \angle F = 60°$, we can conclude that the third angles must also be equal (since the sum of angles in a triangle is always 180°):
  $$\angle B = \angle E = 75°$$
- Since two angles of $\triangle ABC$ are equal to two angles of $\triangle DEF$, by the *AA* criterion, the triangles are similar: $\triangle ABC \sim \triangle DEF$

Similar Figures and Indirect Measurement

Indirect measurement is a method of using proportions and similar figures to find distances or lengths that are difficult to measure directly. This technique often involves the use of similar triangles or other similar shapes.

Key Concepts

1. **Similar Figures**: Figures that have the same shape but different sizes. Corresponding angles are equal, and corresponding side lengths are proportional.
2. **Proportions**: Ratios of corresponding side lengths of similar figures are equal.

Example

Suppose you want to find the height of a tree, but it's difficult to measure directly. You can use the concept of similar triangles to find the height indirectly.

Scenario:
- You place a stick of known height in the ground.
- You measure the length of the shadow cast by the stick and the tree.
- Using the properties of similar triangles, you can set up a proportion to find the height of the tree.

Given:

Height of the stick (h_{stick}): 2 meters

Length of the shadow of the stick (s_{stick}): 1.5 meters

Length of the shadow of the tree (s_{tree}): 6 meters

Find: Height of the tree (h_{tree})

Solution:
1. Set up the proportion: Since the triangles formed by the stick and its shadow, and the tree and its shadow are similar:
$$\frac{(h_{stick})}{(s_{stick})} = \frac{(h_{tree})}{(s_{tree})}$$
2. Plug in the known values:
$$\frac{2}{1.5} = \frac{(h_{tree})}{6}$$
3. Solve for h_{tree}: $(h_{tree}) = \frac{2 \times 6}{1.5} = \frac{12}{6} = 8$ meters

So, the height of the tree is 8 meters.

Congruent Figures

Congruent figures are shapes that are exactly the same in shape and size. This means that one figure can be transformed into the other through a series of rotations, translations, reflections, or a combination of these, without any resizing.

Key Properties of Congruent Figures:

1. **Corresponding Angles**: All corresponding angles are equal.
2. **Corresponding Sides**: All corresponding sides are equal in length.

If two figures A and B are congruent, we write: $A \cong B$

Examples of Congruent Figures:

- **Congruent Triangles:** If two triangles have the same size and shape, they are congruent. For example, if $\triangle ABC$ and $\triangle DEF$ have $AB = DE, BC = EF, CA = FD$ and corresponding angles $\angle A = \angle D, \angle B = \angle E, \angle C = \angle F$ then $\triangle ABC \cong \triangle DEF$.
- **Congruent Rectangles:** If two rectangles have the same length and width, they are congruent.
- **Congruent Circles:** If two circles have the same radius, they are congruent.

Example:

If the two following triangles are congruent, find the missing values ($a, c, e, \angle C, \angle F$ and $\angle E$).

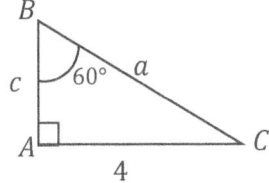 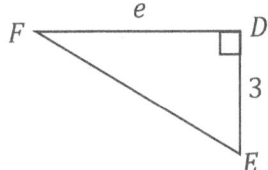

Solution:

Since these two triangles are congruent, we have $AB = DE, BC = EF, CA = FD$ and $\angle A = \angle D, \angle B = \angle E, \angle C = \angle F$ so:

- Finding c and e: $c = AB = DE = 3$ and $e = FD = AC = 4$
- Finding a: The triangles are right by using the Pythagorean theorem:

$BC^2 = AB^2 + AC^2 \rightarrow a^2 = 3^2 + 4^2 = 9 + 16 = 25 \rightarrow a = \sqrt{25} \rightarrow a = 5$

- Finding $\angle C$: The sum of the interior angles of a triangle is equal to 180 degrees:

$\angle C + \angle B + \angle A = 180° \rightarrow \angle C + 60° + 90° = 180° \rightarrow \angle C = 30°$

- Finding $\angle F$ and $\angle E$: $\angle F = \angle C = 30°$ and $\angle E = \angle B = 60°$

So, $a = 5, c = 3, e = 4, \angle C = 30°, \angle F = 30°$ and $\angle E = 60°$.

Congruence Criteria of Triangles

There are several ways to prove that two triangles are congruent, known as congruence criteria:

1. **Side-Side-Side (SSS) Congruence**: If all three sides of one triangle are equal to the corresponding three sides of another triangle, the triangles are congruent:
$$AB = DE, BC = EF, AC = DF \Rightarrow \triangle ABC \cong \triangle DEF$$

2. **Side-Angle-Side (SAS) Congruence**: If two sides and the included angle of one triangle are equal to the corresponding two sides and the included angle of another triangle, the triangles are congruent.
$$AB = DE, \angle B = \angle E, BC = EF \Rightarrow \triangle ABC \cong \triangle DEF$$

3. **Angle-Side-Angle (ASA) Congruence**: If two angles and the included side of one triangle are equal to the corresponding two angles and the included side of another triangle, the triangles are congruent.
$$\angle A = \angle D, AB = DE, \angle B = \angle E \Rightarrow \triangle ABC \cong \triangle DEF$$

4. **Angle-Angle-Side (AAS) Congruence**: If two angles and a non-included side of one triangle are equal to the corresponding two angles and the non-included side of another triangle, the triangles are congruent.
$$\angle A = \angle D, \angle B = \angle E, AC = DF \Rightarrow \triangle ABC \cong \triangle DEF$$

Examples:

1) Determine if two given triangles are congruent and identify the congruence criterion.
   - $\triangle ABC$ with sides $AB = 5, BC = 6, CA = 7$
   - $\triangle DEF$ with sides $DE = 5, EF = 6, FD = 7$

   Solution:

   Check the corresponding sides:
   - $DE = AB = 5$
   - $EF = BC = 6$
   - $FD = CA = 7$

   Since all three pairs of corresponding sides are equal, the triangles are congruent by the SSS $(Side - Side - Side)$ congruence criterion.

2) Prove that $\triangle ABC \cong \triangle DEF$ given the following information:
   $\angle A = \angle D, BC = EF$ and $\angle B = \angle E$

   Solution:
   - We have two pairs of equal angles: $\angle A = \angle D$ and $\angle B = \angle E$.
   - We also have the non-included side $BC = EF$.
   - We conclude that: $\triangle ABC \cong \triangle DEF$ by AAS.

Word Problems

Steps for Congruence Problems:

1. **Read the Problem Carefully**: Identify the given information and what needs to be proven or found.

2. **Identify the Figures and Criteria**:
   - Determine which geometric figures are involved (usually triangles).
   - Identify the appropriate similarity criterion (*AA, SSS or SAS*) or congruence criterion (*SSS, SAS, ASA, AAS*).

3. **Write Down the Given Information**: List all the known side lengths and angle measures.

4. **Apply the Similarity or Congruence Criterion**:
   - Use the identified criterion to compare the corresponding parts of the figures.
   - Verify if all conditions for the selected criterion are met.

5. **Set Up Proportion for Similarity or Write the Congruence Statement**:
   - If the figures are similar, set up proportions to compare corresponding sides.
   - If the figures are congruent, write the congruence statement (e.g., $\triangle ABC \cong \triangle DEF$).

6. **Solve for Unknowns**:
   - Use the proportions to solve the unknown side lengths or angles.
   - Use the congruence to find the required measurements or prove the given statement.

Example:

Emma and Jack are each building a triangular shelf. Emma's shelf has two sides of 5 inches and 7 inches with an included angle of 45 degrees. Jack's shelf has two sides of 5 inches and 7 inches with an included angle of 45 degrees. Are the shelves congruent?

Solution: To determine if the shelves are congruent, we'll use the *SAS* (*Side − Angle − Side*) congruence criterion.

1. Identify the given information:
   - Emma's shelf: 5 inches, 7 inches, and included angle 45°.
   - Jack's shelf: 5 inches, 7 inches, and included angle 45°.
2. Compare the sides and angle: Both shelves have two sides that are 5 inches and 7 inches, and the included angle is 45°.
3. Apply the *SAS* criterion: Since both sides and the included angle are equal, by the *SAS* criterion, the shelves are congruent.

Worksheets

📎 Similar Figures

Find the missing values (x, y, z, t) in the similar shapes:

1)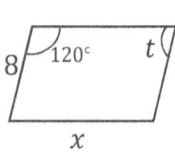

2) The triangles are isosceles.

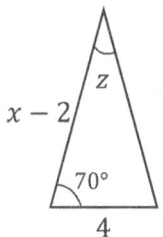 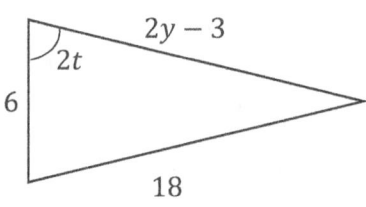

3)

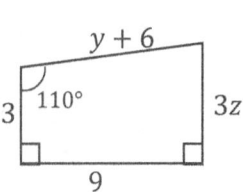 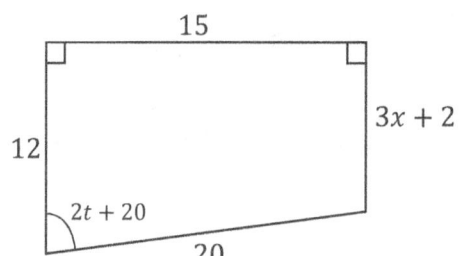

4) The scale factor is $\frac{4}{7}$.

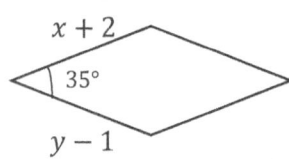 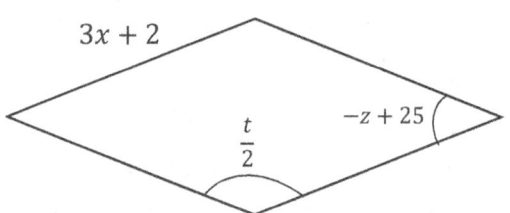

📎 Ratio of Area and Volume in similar Figures

Do the following problems.
1) Two similar triangles have a scale factor of 2. If the area of the smaller triangle is 10 square units, what is the area of the larger triangle?
2) Two similar circles have radii in the ratio of 1: 6. If the area of the smaller circle is 12.5 square units, what is the area of the larger circle?

3) Two similar cones have heights in the ratio of 1: 4. If the volume of the larger cone is 256 cubic units, what is the volume of the smaller cone?
4) We have two squares, one of which has a diagonal of 32 cm, and the other has an area of 36 square cm. Find the scale factor.
5) We have two similar rectangles with a scale factor of 3: 8. If the area of one is $2x + 1$ square cm and the area of the other is $5x - 3$ square cm, find the value of x.

Similarity Criteria for Triangles
Solve the problems.
1) Triangle XYZ is similar to triangle ABC. If the sides of triangle XYZ are 5 cm, 12 cm, and 13 cm, and the shortest side of triangle ABC is 10 cm, find the lengths of the other two sides of triangle ABC.

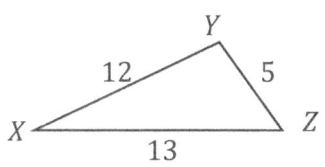
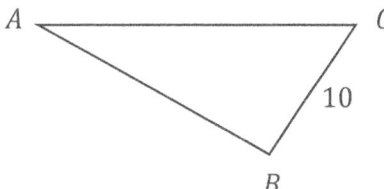

2) In $\triangle PRQ$ and $\triangle SUT$, $\angle P = 60°$ and $\angle Q = 50°$. Also, $\angle S = 60°$ and $\angle U = 70°$. Show that $\triangle PQR \sim \triangle STU$.

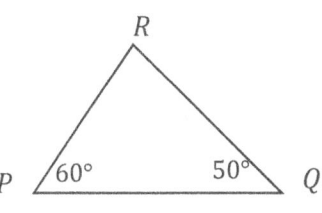
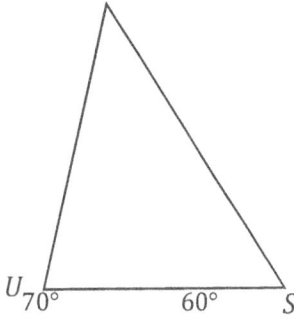

3) Prove that triangles formed by the diagonals of a parallelogram are similar.
4) In right triangle $\triangle ABC$ with $\angle C = 90°$, altitude CD is drawn to the hypotenuse AB. Prove: $\triangle BDC \sim \triangle BCA$

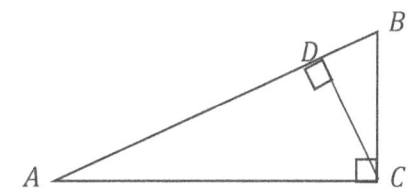

5) Quadrilateral $ABCD$ has $AB \parallel CD$. Diagonals AC and BD intersect at O. Proof:
- $\triangle AOB \sim \triangle COD$
- $AO \cdot DO = BO \cdot CO$

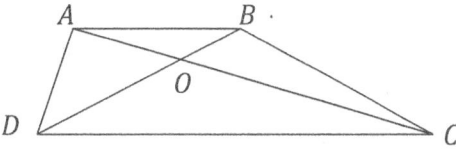

Similar Figures and Indirect Measurement

Find the answers.

1) A 1.8-meter-tall man casts a shadow 2.4 meters long. At the same time, a nearby tree casts a shadow of 15 meters. Find the height of the tree.

2) A flagpole casts a shadow of 20 meters, and a nearby 1.5-meter-tall person casts a shadow of 3 meters. Find the height of the flagpole.

3) A model car is 20 cm long, and the actual car is 2 meters long. What is the scale factor from the model to the actual car?

4) A large rectangular picture frame and a small rectangular picture frame are similar. The dimensions of the larger frame are 120 cm by 150 cm, and the dimensions of the smaller frame are 40 cm by 50 cm. Find the ratio of the areas and the ratio of the perimeters of the two frames.

5) A map has a scale of $1:50,000$, and the distance between two cities on the map is 8 cm. Using similar triangles, find the actual distance between the two cities in kilometers.

Congruent Figures

Do the problems.

1) Triangle $\triangle ABC$ is congruent to triangle $\triangle DEF$. If $\angle A = 45°, \angle B = 60°$, and $AB = 5$ cm, find the measures of $\angle D, \angle E$, and DE.

2) If a rectangle $ABCD$ and a rectangle $PQRS$ are congruent, and the length of $AB = 8$ cm, the width of $BC = 5$ cm, and the area of the rectangle is given, calculate the area of rectangle $PQRS$.

3) Two circles have congruent radio. If the radius of the first circle is 6 cm, and the center-to-center distance between the circles is 10 cm, find the x.

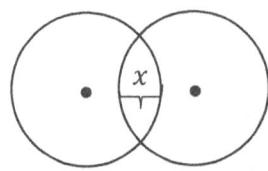

4) Quadrilateral $PQRS$ is congruent to quadrilateral $TUVW$. If $PQ = 7$ cm, $QR = 6$ cm, $RS = 5$ cm, $SP = 4$ cm, and $\angle PQR = 90°$, find the lengths of TU, UV, VW, WT, and the measure of $\angle TUV$.

5) Two congruent trapezoids have bases of lengths 10 cm and 6 cm, and height 4 cm. Find the lengths of the non-parallel sides and the area of one trapezoid.

Congruence Criteria of Triangles

Solve the problems.

1) If in rectangle $ABCD$ we draw the diagonal BD, prove that $\triangle ABD \cong \triangle BCD$.

2) In $\triangle ABC$, $\angle C = \angle B$ and points D and E are midpoints of AB and AC respectively. Prove $\triangle ECB \cong \triangle DBC$.

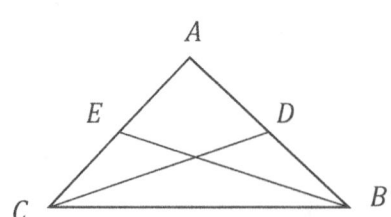

3) In △ PQR, PS is the perpendicular bisector of QR. Prove △ PQS ≅ △ PRS.
4) In the figure below O is the center of both circles, prove that the two triangles △ ABO and △ OCD are congruent.

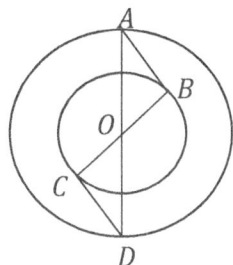

5) The triangle △ ABC is isosceles, with AB = AC. BH is the altitude from vertex B, and CH' is the altitude from vertex C. Why is BH equal to CH'?

Word Problems

Find the answers.

1) Two playground slides are congruent. One slide has a length of 15 feet and an angle of elevation of 30 degrees. What is the length and angle of elevation of the other slide?
2) Two buildings are similar. The height of the first building is 80 meters and the width at the base is 20 meters. If the height of the second building is 120 meters, find the width of the second building.
3) A model airplane is built to a scale where 1 inch represents 2 feet. If the model's wingspan is 6 inches, what is the wingspan of the actual airplane in feet?
4) Two similar gardens have areas of 48 square meters and 108 square meters. If the width of the smaller garden is 6 meters, what is the width of the larger garden?
5) A small TV screen has a diagonal of 20 inches, and a larger TV screen is similar to the smaller one with a diagonal of 50 inches. If the area of the smaller screen is 160 square inches, what is the area of the larger screen?
6) In a triangular roof design, two congruent right triangles are placed next to each other. The length of the hypotenuse is 12 feet, and one leg is 9 feet. What is the length of the other leg?
7) You are tiling a rectangular floor with square tiles. Each tile is 30 cm by 30 cm, and all tiles are congruent. If the floor is 3 meters by 4 meters, how many tiles do you need to cover the entire floor?
8) A triangular roof section has a height of 10 feet and a base of 12 feet. Another triangular roof section is congruent to the first, but its base is 18 feet. If the base of the second roof section is enlarged, what is the height of the second roof section?
9) A 1.5-meter-tall person is standing 20 meters away from a building. The angle of elevation from the person's eyes to the top of the building is 45°. Using the concept of similar triangles, estimate the height of the building.
10) Two maps of the same region use different scales. Map 1 uses a scale of 1: 50,000, and Map 2 uses a scale of 1: 100,000. On Map 1, the distance between two cities is measured as 5 cm. Find the distance between the same two cities on Map 2 using the concept of similar figures and proportionality.

www.mathnotion.com

Answer of Worksheets

Similar Figures
1) $x = 10, y = 60°, z = 120°$ and $t = 60°$
2) $x = 14, y = \frac{21}{2}, z = 40°$ and $t = 35°$
3) $x = 1, y = 6, z = \frac{12}{5}$ and $t = 25°$
4) $x = \frac{6}{5}, y = \frac{21}{5}, z = -10°, t = 290°$

Ratio of Area and Volume in similar Figures
1) 40 square units
2) 450 square units
3) 4 cubic units
4) $\frac{8\sqrt{2}}{3}$
5) $-\frac{91}{83}$

Similarity Criteria for Triangles
1) 24 and 26 cm.
2) Find the missing angles in both triangles:
   - In $\triangle PQR$: Given: $\angle P = 60°$ and $\angle Q = 50°$. Using the Angle Sum Property of a triangle ($\angle P + \angle Q + \angle R = 180°$): $\angle R = 180° - \angle P - \angle Q = 180° - 60° - 50° = 70°$.
   - In $\triangle STU$: Given: $\angle S = 60°$ and $\angle U = 70°$. Using the Angle Sum Property of a triangle: $\angle T = 180° - \angle S - \angle U = 180° - 60° - 70° = 50°$.

   Compare corresponding angles:
   - $\triangle PQR$: $\angle P = 60°, \angle Q = 50°, \angle R = 70°$
   - $\triangle STU$: $\angle S = 60°, \angle T = 50°, \angle U = 70°$

   Conclusion (AA similarity criterion): Since two angles of one triangle are equal to two angles of the other triangle, by the AA (Angle-Angle) Similarity Criterion, the triangles are similar. Thus, $\triangle PQR \sim \triangle STU$.

3) Let $ABCD$ be a parallelogram with diagonals AC and BD intersecting at point O. The triangles $\triangle OAB, \triangle OBC, \triangle OCD$, and $\triangle ODA$ are formed.

   Proof using AA similarity criterion:
   - Opposite angles in a parallelogram are equal: $\angle OAB = \angle OCD$ and $\angle OBC = \angle ODA$.
   - Vertically opposite angles are equal: $\angle AOB = \angle DOC$ and $\angle BOC = \angle AOD$.
   - Since each pair of triangles has two corresponding equal angles, they satisfy the AA Similarity Criterion, which means: $\triangle OAB \sim \triangle OCD$ and $\triangle OBC \sim \triangle ODA$.

4) In $\triangle BDC$ and $\triangle BCA$:
   - Both have a right angle: $\angle BDC = 90°$ (since CD is an altitude), $\angle BCA = 90°$ (given).
   - They share $\angle B$.

   Therefore, by the AA (Angle-Angle) Similarity Criterion, the triangles are similar.

5) Step 1: Prove $\triangle AOB \sim \triangle COD$:
   Alternate Angles:

www.mathnotion.com

- $\angle OAB = \angle OCD$ (alternate angles, since $AB \parallel CD$ and AC is a transversal).
- $\angle OBA = \angle ODC$ (alternate angles, since $AB \parallel CD$ and BD is a transversal).

vertical angles:
- $\angle AOB = \angle COD$ (vertically opposite angles).

AA Similarity Criterion:
- Since two pairs of corresponding angles are equal ($\angle OAB = \angle OCD$ and $\angle OBA = \angle ODC$), by the AA (Angle-Angle) Similarity Criterion, the triangles are similar. Thus, $\triangle AOB \sim \triangle COD$.

Step 2: Prove $AO \cdot DO = BO \cdot CO$: From the similarity $\triangle AOB \sim \triangle COD$, the corresponding sides are proportional: $\frac{AO}{CO} = \frac{BO}{DO} = \frac{AB}{CD}$. Cross-multiplying the first two ratios gives: $AO \cdot DO = BO \cdot CO$

Similar Figures and Indirect Measurement
1) 11.25 meters
2) 10 meters
3) $1:10$
4) Ratio of the area: $9:1$ and ratio of the perimeters: $3:1$
5) 4 kilometers

Congruent Figures
1) $\angle D = 45°, \angle E = 60°,$ and $DE = 5\ cm$
2) 40 square cm
3) 2 cm
4) $TU = 7\ cm, UV = 6\ cm, VW = 5\ cm, WT = 4\ cm$, and the measure of $\angle TUV = 90°$
5) The lengths of the non-parallel sides $= 2\sqrt{5}$ and the area of one trapezoid $= 32$ sq cm

Congruence Criteria of Triangles
1) In rectangle $ABCD$, diagonal BD is drawn, we will prove $\triangle ABD \cong \triangle BCD$
   Common side: Both triangles share side BD (the diagonal). Thus, $BD = BD$ (reflexive property).
   Equal sides: $AB = CD$ (opposite sides of a rectangle) and $AD = BC$ (opposite sides of a rectangle).
   Apply the SSS congruence criterion:

   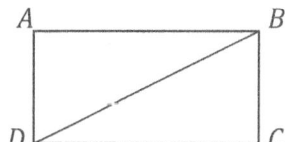

   - Side (S): $BD = BD$ (common side)
   - Side (S): $AB = CD$ (given)
   - Side (S): $AD = BC$ (given)

   By the SSS (Side-Side-Side) congruence criterion, $\triangle ABD \cong \triangle BCD$.

2) Step 1: Analyze information given:
   - Angles: $\angle B = \angle C$ (given). This implies $\triangle ABC$ is isosceles with $AB = AC$ (sides opposite equal angles are equal).
   - Midpoints: D is the midpoint of AB $\Rightarrow AD = DB = \frac{1}{2} AB$. E is the midpoint of $AC \Rightarrow AE = EC = \frac{1}{2} AC$. Since $AB = AC$, their halves are also equal: $DB = EC$.

   Step 2: Identify corresponding parts of $\triangle ECB$ and $\triangle DBC$.
   - Common side: BC is shared by both triangles
   - Equal side: $DB = EC$ (from Step 1)
   - Equal angles: $\angle B = \angle C$ (given)

   Step 3: Apply the SAS congruence criterion:

- Side (S): $DB = EC$ (proven in Step 1)
- Angle (A): $\angle B = \angle C$ (given)
- Side (S): $BC = CB$ (common side)

Thus, by SAS (Side-Angle-Side) congruence, $\triangle ECB \cong \triangle DBC$.

3) Step 1: Understand the information given:
Perpendicular bisector definition: PS is perpendicular to $QR \Rightarrow \angle PSR = \angle PSQ = 90°$. In addition, PS bisects $QR \Rightarrow QS = SR$.
Step 2: Identify corresponding parts of $\triangle PQS$ and $\triangle PRS$.
- Common side: PS is shared by both triangles
- Equal side: $QS = SR$ (since PS bisects QR)
- Equal angles: $\angle PSQ = \angle PSR = 90°$ (since $PS \perp QR$)

Step 3: Apply the SAS congruence criterion:
- Side (S): $QS = SR$ (given by bisector property)
- Angle (A): $\angle PSQ = \angle PSR = 90°$ (given by perpendicularity)
- Side (S): $PS = PS$ (common side)

Thus, by SAS (Side-Angle-Side) congruence, $\triangle PQS \cong \triangle PRS$.

4) Step 1: Identify corresponding parts of $\triangle ABO$ and $\triangle OCD$:
- Equal side: $OB = OC$ (since both are the radii of the smaller circle)
- Equal side: $OA = OD$ (since both are the radii of the larger circle)
- Equal angle: $\angle AOB = \angle DOC$ (vertically opposite angles)

Step 2: Apply the SAS congruence criterion:
- Side (S): $OB = OC$ (the radii of the smaller circle)
- Side (A): $\angle AOB = \angle DOC$ (vertically opposite angles)
- Side (S): $OA = OD$ (the radii of the larger circle)

Thus, by SAS (Side-Angle-Side) congruence, $\triangle ABO \cong \triangle OCD$

5) Step 1: Analyze the right triangles formed by the altitudes:
$\triangle BHC$ and $\triangle CH'B$ are both right triangles: $BH \perp AC \Rightarrow \angle BHC = 90°$ and $CH' \perp AB \Rightarrow \angle CH'B = 90°$
Step 2: Prove $\triangle BHC \cong \triangle CH'B$:
- $\angle BHC = \angle CH'B = 90°$ (Both are right angles by definition of altitudes).
- $\angle HCB = \angle H'BC$ (since the triangle is isosceles and $AB = AC$).
- $BC = CB$ (Common side to both triangles).

Thus, $\triangle BHC \cong \triangle CH'B$ by AAS.
Step 3: Conclude $BH = CH'$:
Since corresponding parts of congruent triangles are equal, the altitudes BH and CH' are equal.

Word Problems
1) The length of the other side: 15 feet and angle of elevation of the other slide: 30 degrees
2) 30 meters
3) 12 feet
4) 9 meters
5) 1,000 square inches
6) $\sqrt{63}$ feet
7) 134 tiles
8) 15 feet
9) 21.5 meters
10) 2.5 cm

Chapter 13: Geometry and Solid Figures

Topics that you'll learn in this chapter:

✓ **Angles**
- Angles of Triangles and Quadrilateral
- Interior and Exterior Angles of Polygons
- Angle of Circles

✓ **Area and Perimeter**
- Pythagorean Theorem
- Area of Compound Figures
- Perimeter of Compound Figures

✓ **Volume and Surface Area**
- Front, Side and Top of Three-Dimensional Figures
- Cubes
- Rectangular Prism
- Cylinder
- Pyramids
- Cone

✓ Word Problems

✓ Worksheets

✓ Answer of Worksheets

Angles of Triangles and Quadrilateral

Triangles:

The triangle is a three-sided polygon. The sum of the interior angles in any triangle is always 180°. Here are the different types of triangles based on their angles:

- **Acute Triangle**: All three angles are less than 90°.
- **Right Triangle**: One angle is exactly 90°.
- **Obtuse Triangle**: One angle is greater than 90°.

Quadrilaterals:

A quadrilateral is a four-sided polygon. The sum of the interior angles in any quadrilateral is always 360°. Here are some common types of quadrilaterals based on their angles:

- **Square**: All four angles are 90°.
- **Rectangle**: All four angles are 90°, but opposite sides are equal.
- **Parallelogram**: Opposite angles are equal, and opposite sides are parallel and equal.
- **Rhombus**: All sides are equal, but the angles are not necessarily 90°. Opposite angles are equal.
- **Trapezoid (or Trapezium)**: At least one pair of opposite sides are parallel. Angles on the same side of the non-parallel sides are supplementary (add up to 180°).

Example:

Find the missing values in the following shapes (the dashed line is a bisector):

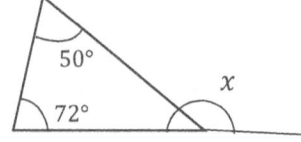 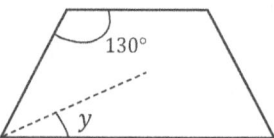

Solution:

Finding x: The sum of the interior angles in any triangle is always 180°:

The third interior angle $= 180° - (50° + 72°) = 58°$

$x = 180° - 58° = 122°$

Finding y: Angles on the same side of the non-parallel sides are supplementary:

$180° - (130°) = 50°$ and the bisector divides the angle into two equal parts: $y = 50° \div 2 = 25°$

Interior and Exterior Angles of Polygons

Interior Angles:

Interior angles are the angles inside the polygon, formed by two adjacent sides. The sum of the interior angles of a polygon depends on the number of sides (n) the polygon has. Here's the formula to calculate the sum of the interior angles:

$$Sum\ of\ interior\ angles = (n - 2) \times 180°$$

Exterior Angles:

Exterior angles are the angles formed between one side of the polygon and the extension of an adjacent side. The sum of the exterior angles of any polygon is always 360°.

☑ For regular polygons (where all sides and angles are equal), the measure of each exterior angle can be calculated using the formula:

$$Each\ exterior\ angle = \frac{360°}{n}$$

Examples:

1) Find the sum of interior angles of pentagon.

   Solution:

   A pentagon has 5 sides so the sum of interior angles of pentagon= $(5 - 2) \times 180° = 540°$

2) Find each interior angle of a regular octagon.

   Solution:

   Find the sum of interior angles of an octagon: $(8 - 2) \times 180° = 1,080°$

   Find each interior angle: $1,080° \div 8 = 135°$

3) If each exterior angle of a regular polygon measures 30 degrees, determine the number of its sides.

   Solution:
   1. The formula for finding each exterior angle of a regular polygon is: $\theta = \frac{360°}{n}$
   2. Substitute the given exterior angle into the formula: $30° = \frac{360°}{n}$
   3. Solve for n: $30° = \frac{360°}{n} \rightarrow 30° \times n = 360° \rightarrow n = \frac{360°}{30°} = 12$

   So, the number of sides in the regular polygon is 12.

Angle of Circle

In geometry, a circle is a 2D shape where all points are equidistant from a central point. The angle around a circle, also known as the central angle, is measured in degrees or radians.

Key Points:

- **Full Circle:** The total angle around a circle is 360 degrees or 2π radians.
- **Half Circle:** A semicircle has an angle of 180 degrees or π radians.
- **Quarter Circle:** A quarter circle, like a right angle, has an angle of 90 degrees or $\frac{\pi}{2}$.

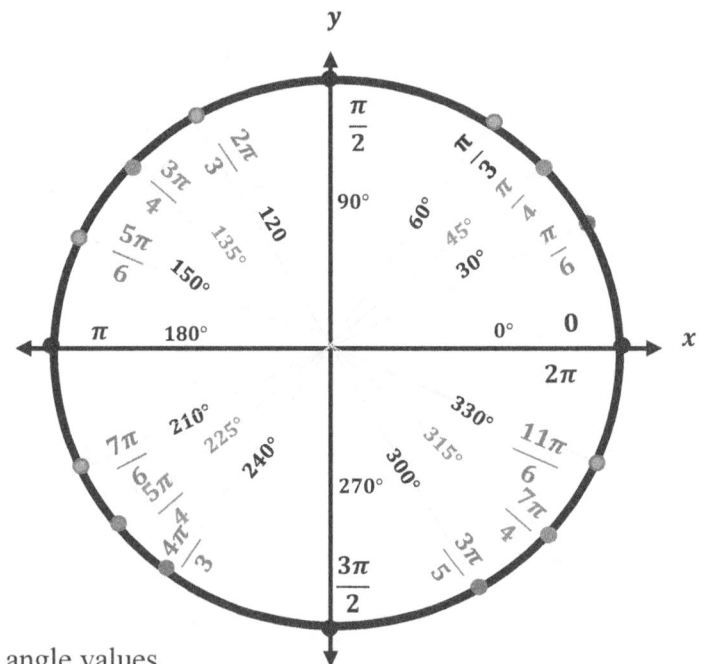

Example:

Find the missing angle values.

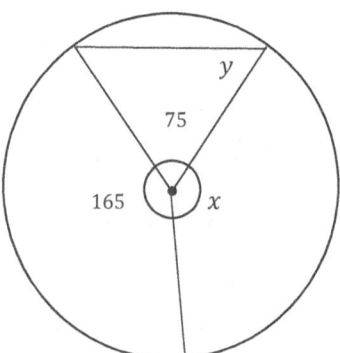

Solution:

- The total angle around a circle is 360°, so we have:
$165° + 75° + x = 360°$
$240° + x = 360°$
$x = 360° - 240° = 120°$

- The angle y belongs to an isosceles triangle (because the radii of circle are equal) and the sum of all three angles of triangles is 180°:
$y + y + 75° = 180°$
$2y + 75° = 180°$
$2y = 180° - 75° = 105°$
$y = 52.5°$

www.mathnotion.com

Pythagorean Theorem

The Pythagorean Theorem is a fundamental principle in geometry that describes the relationship between the sides of a right triangle. It is named after the ancient Greek mathematician Pythagoras.

Key Concepts

1. **Right Triangle**: A triangle with one angle that is exactly 90 degrees.
2. **Hypotenuse**: The side opposite the right angle, which is the longest side of the right triangle.
3. **Legs**: The two sides that form the right angle.

Pythagorean Theorem Formula

The Pythagorean Theorem states that:

$$a^2 + b^2 = c^2$$

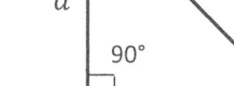

where:

- a and b are the lengths of the legs of the right triangle.
- c is the length of the hypotenuse.

Examples:

1) Suppose we have a right triangle with legs of lengths 3 cm and 4 cm, and we want to find the hypotenuse length.
   Solution:

   1. Identify the lengths of the legs and substitute the values into the formula:
      $a = 3\ cm\ and\ b = 4\ cm\ \rightarrow\ a^2 + b^2 = c^2\ \rightarrow\ 3^2 + 4^2 = c^2$

   2. Calculate the square of the legs and find c: $9 + 16 = c^2\ \rightarrow\ 25 = c^2\ \rightarrow\ c = \sqrt{25} = 5\ cm$

2) Find the area of the following rectangle:

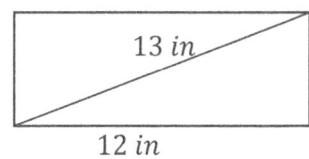

   Solution

   To calculate the area of a rectangle, we need its width and length. We can determine the width by using the Pythagorean theorem:

   Here $c = 13\ in, b = 12\ in\ \rightarrow\ a^2 + 12^2 = 13^2\ \rightarrow\ a^2 = 169 - 144 = 25\ \rightarrow\ a = 5\ in$

   The width is $5\ in$ and the length is $12\ in$, so the area will be $12 \times 5 = 60\ in^2$.

Area of Compound Figures

The area of compound figures is calculated by breaking them down into simpler shapes (like rectangles, triangles, circles, etc.), then finding the area of each shape, and finally adding these areas together.

Steps to Find the Area of Compound Figures:

1. **Identify the Simple Shapes:** Look at the compound figure and identify the simple shapes that make it up (e.g., rectangles, triangles, circles).
2. **Divide the Compound Figure:** Split the compound figure into these identified simple shapes.
3. **Calculate the Area of Each Shape:** Use the appropriate formula for each shape:
   - Rectangle: $Area = length \times width$
   - Triangle: $Area = \frac{1}{2} \times base \times height$
   - Circle: $Area = \pi \times radius^2$
   - Trapezoid: $Area = \frac{1}{2} \times (base_1 + base_2) \times height$

Add the Areas Together: Once you have the areas of all the simple shapes, add them together to get the total area of the compound figure. ($\pi \approx 3.14$)

Example:

Find the area of the following compound shape:

Solution:

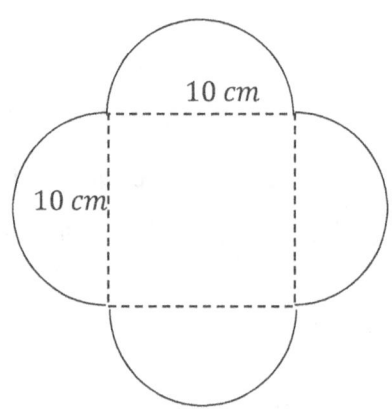

1. Identify the simple shapes: This shape consists of 4 semicircles and a square, or it consists of two full circles and a square.
2. Calculate the area of each shape:

   Area of square: $Area = side \times side = side^2 = 10 \times 10 = 100 \ cm^2$

   Area of circle $= Area = \pi \times radius^2 = \pi \times 5^2 = 25\pi \approx 78.5 \ cm^2$

   Area of 2 circles $= Area = 2 \times \pi \times radius^2 \approx 2 \times 78.5 = 157 \ cm^2$

3. Add the area together:

   $Area \ of \ square + Area \ of \ 2 \ circles \approx 100 + 157 = 257 \ cm^2$

www.mathnotion.com

Perimeter of Compound Shapes

To find the perimeter of compound shapes, you need to sum up the lengths of all the outer sides of the shape. Here are the steps:

Steps to Find the Perimeter of Compound Figures:

1. **Identify the Outer Edges:** Look at the compound shape and identify all the outer edges that make up the perimeter.
2. **Calculate the Length of Each Edge:** Measure or find the length of each outer edge. In this process, you may need the perimeter formulas for some shapes:
   - Rectangle: $Perimeter = 2 \times (length + width)$
   - Square: $Perimeter = 4 \times Side$
   - Circle: $Perimeter = 2 \times r \times \pi$
3. **Add the Lengths Together:** Sum the lengths of all these outer edges to get the total perimeter.

Example:

Find the perimeter of the following compound shape: ($\pi \approx 3.14$)

Solution:

1. Identify the outer edges: There are 3 segments (red parts) and a semicircle (blue part):
2. Calculate the length of each edge:
   - The length of 3 segments are $40\ ft$, $20\ ft$ and $40\ ft$.
   - The length of semicircle: $\frac{1}{2} \times 2 \times r \times \pi = r \times \pi = 10 \times 3.14 = 31.4\ ft$
3. Add the lengths together: Sum the lengths of all these outer edges to get the total perimeter: Total perimeter: $40 + 20 + 40 + 31.4 = 131.4\ ft$

Front, Side and Top of Three-Dimensional Figures

When working with three-dimensional (3D) figures, we often describe them using three views: the front view, the side view, and the top view. These views help us understand the shape and structure of the 3D figure by looking at it from different perspectives.

1. **Front View:** This is what you see when you look directly at the front of the 3D figure. It shows the height and width of the figure but not the depth.

2. **Side View:** This is what you see when you look directly at one of the sides of the 3D figure. It shows the height and depth but not the width.

3. **Top View:** This is what you see when you look directly down at the top of the 3D figure. It shows the width and depth but not the height.

Let's consider rectangular prism as an example:

- **Front View:** You'll see a rectangle that represents the height and width of the prism.
- **Side View:** You'll see another rectangle that represents the height and depth.
- **Top View:** You'll see a rectangle representing the width and depth.

Example:

Identify how the three-dimensional shapes below are seen from the front, side, and top views:

a) b) c)

Solution

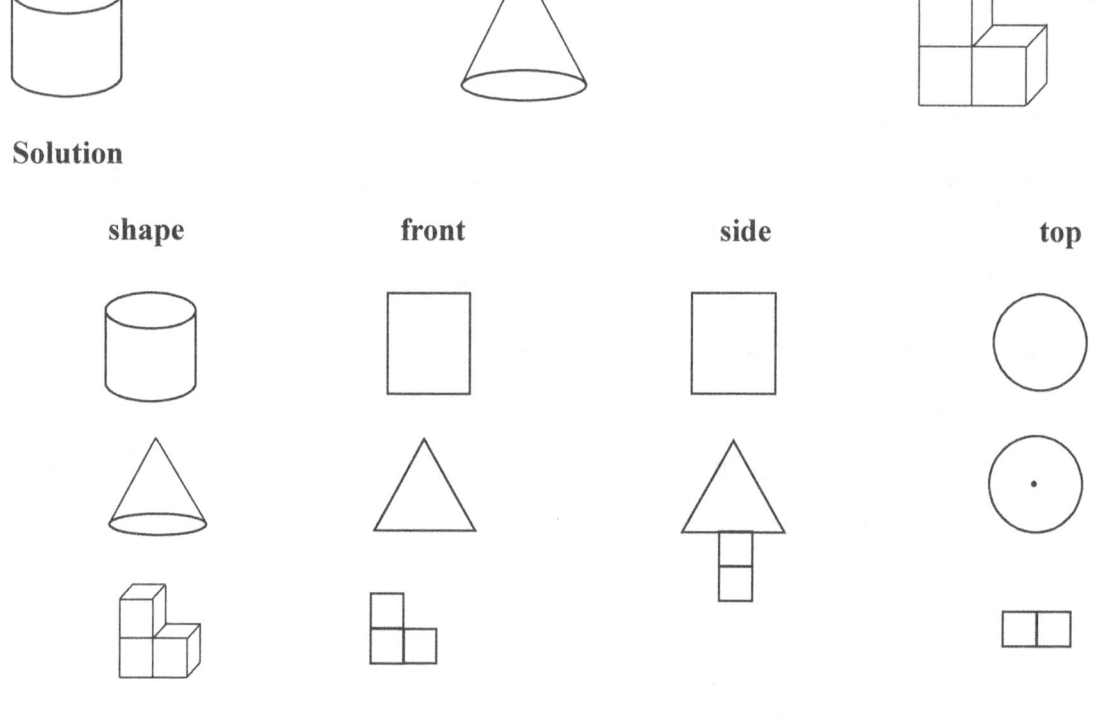

Cubes

A cube is a three-dimensional geometric shape with six equal square faces. It is a type of rectangular prism where all the sides are equal in length.

Key Properties of a Cube:
1. **Faces**: A cube has 6 faces, and each face is square.
2. **Edges**: A cube has 12 edges, and all edges are equal in length.
3. **Vertices**: A cube has 8 vertices (corners).

Formula for a Cube:
- **Surface Area** (Total area): The surface area S of a cube can be calculated by finding the area of one face and multiplying it by 6 (since there are 6 faces):
$$S = 6a^2$$
where a is the length of an edge of the cube.
- **Volume:** The volume V of a cube can be calculated by finding the volume of the space inside the cube:
$$V = a^3$$

Examples:

1) A cube has an edge length of 0.5 m. What is its volume and total surface area?

   Solution:
   - The volume of a cube is given by:
   $$V = side^3 = 0.5^3 = 0.5 \times 0.5 \times 0.5 = 0.125 \text{ cube meters.}$$
   - The surface area of a cube is given by:
   $$A = 6 \times side^2 = 6 \times 0.5^2 = 6 \times 0.5 \times 0.5 = 1.5 \text{ square meters.}$$

2) If we cut a small cube with a side of 1 centimeter from one corner of a cube with a side of 10 centimeters, find the volume of the new shape.

   Solution:
   1. Calculate the volume of the original cube: The original cube has a side length of 10 centimeters:
   $$V_{original} = side^3 = 10^3 = 10 \times 10 \times 10 = 1,000 \text{ cube centimeters}$$
   2. Calculate the volume of the small cube: The small cube has a side length of 1 centimeter:
   $$V_{small} = side^3 = 1^3 = 1 \times 1 \times 1 = 1 \text{ cube centimeters}$$
   3. Subtract the volume of the small cube from the original cube:
   $$V_{new} = V_{original} - V_{small} = 1,000 - 1 = 999 \text{ cubic centimeters.}$$

Rectangular Prism

A rectangular prism is a three-dimensional geometric shape with six faces, all of which are rectangles. It is also known as a cuboid.

Key Properties of a Rectangular Prism:

1. **Faces**: A rectangular prism has 6 rectangular faces.
2. **Edges**: A rectangular prism has 12 edges.
3. **Vertices**: A rectangular prism has 8 vertices (corners).

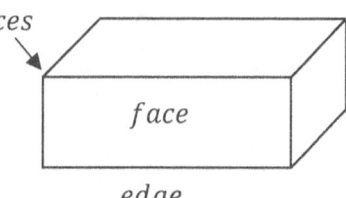

Surface Area:

The surface area S of a rectangular prism can be calculated by finding the area of each face and summing them up. If the dimensions of the prism are length b, width a, and height h, the surface area is given by:

$$S = 2ab + 2bh + 2ah$$

Volume:

The volume V of a rectangular prism can be calculated by multiplying its length, width, and height:

$$V = abh$$

Examples:

1) Find the volume and surface area of a rectangular prism with a length of 6.2 ft, a width of 4 ft, and a height of 5.5 ft.
   Solution:
   - Calculate the volume of rectangular prism: Using the formula: $V = abh = 6.2 \times 4 \times 5.5 = 136.4$ cubic feet.
   - Calculate the surface area of rectangular prism: Using the formula: $S = 2ab + 2bh + 2ah = 2 \times 6.2 \times 4 + 2 \times 4 \times 5.5 + 2 \times 6.2 \times 5.5 = 161.8$ square feet

2) In a rectangular prism with dimensions 12 cm by 18 cm by 20 cm, how many small cubes with dimensions of 2 cm can fit inside?
   Solution:
   1. Calculate the volume of rectangular prism: Using the formula for the volume: $V = abh = 12 \times 18 \times 20 = 4,320 \ cm^3$
   2. Calculate the volume of a cube: Using the formula for the volume of cube: $V = side^3 = 2 \times 2 \times 2 = 8 \ cm^3$
   3. Calculate the number of small cubes that can fit inside the prism:
   $Number \ of \ small \ cubes = \frac{V_{prism}}{V_{cube}} = \frac{4320}{8} = 540$ small cubes

www.mathnotion.com

Cylinder

A cylinder is a three-dimensional geometric shape with two parallel circular bases connected by a curved surface.

Key Properties of a Cylinder

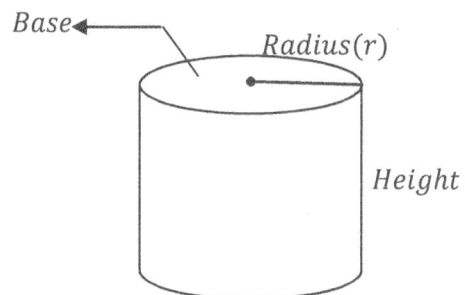

1. **Bases**: A cylinder has two circular bases that are parallel and congruent.

2. **Height (h)**: The perpendicular distance between the two bases.

3. **Radius (r)**: The radius of the circular base.

Surface Area: The surface area S of a cylinder can be calculated by summing the areas of the two bases and the curved surface area (lateral surface area).

- **Base Area**: The area of one circular base is πr^2.
- **Lateral Surface Area**: The area of the curved surface is $2\pi rh$.
- **The total surface area** S is given by:
$$S = 2\pi r^2 + 2\pi rh = 2\pi r(r+h)$$

Volume: The volume V of a cylinder can be calculated by finding the area of the base and multiplying it by the height:
$$V = \pi r^2 h$$

Examples:

1) Calculate the surface area and volume of a cylinder with a radius of $3\ in$ and a height of $5\ in$. ($\pi \approx 3.14$)

   Solution:
   - The total surface area: Using the formula for total surface area:
   $$S = 2\pi r(r+h) = 2\pi \times 3 \times (3+5) = 2\pi \times 3 \times 8 = 48\pi \approx 150.80\ in^2$$
   - The volume: Using the formula for volume of cylinder:
   $$V = \pi r^2 h = \pi(3)^2(5) = 45\pi \approx 141.37\ in^3$$

2) A cylinder has the volume of $1{,}500\ ft^3$ and a height of $5\ ft$. Find its radius. ($\pi \approx 3$)

   Solution:
   Substitute the given values into the volume formula and find r:
   $$V = \pi r^2 h \rightarrow 1{,}500 = 3 \times r^2 \times 5 \rightarrow 1{,}500 = 15r^2$$
   $$r^2 = \frac{1500}{15} = 100 \rightarrow r = \sqrt{100} = 10\ ft$$

Pyramids

A pyramid is a polyhedron with a polygonal base and triangular faces that meet at a common point called the apex. The most common type is the square pyramid, which has a square base and four triangular faces.

Key Properties of a Pyramid

1. **Base**: The bottom polygonal face.
2. **Apex**: The top point where all triangular faces meet.
3. **Height (h)**: The perpendicular distance from the base to the apex.
4. **Slant Height (l)**: The distance from the apex to the midpoint of one of the base's edges.

Surface Area: For a pyramid with a square base:

The surface area S of a pyramid is the sum of the base area and the lateral area (the sum of the areas of the triangular faces).

- **Base Area (B)**: $B = a^2$ (where s is the side length of the base).
- **Lateral Surface Area**: $S_{lateral} = 2al$
- **Total Surface Area**: $S_{total} = B + 2al \rightarrow S_{total} = a^2 + 2al$

where B is the area of the base, a is the side of the base, and l is the slant height.

Volume:

$$V = \frac{1}{3} B \cdot h = \frac{1}{3} a^2 \cdot h$$

Example:

Find the volume and total surface area of the pyramid. (all units are cm)

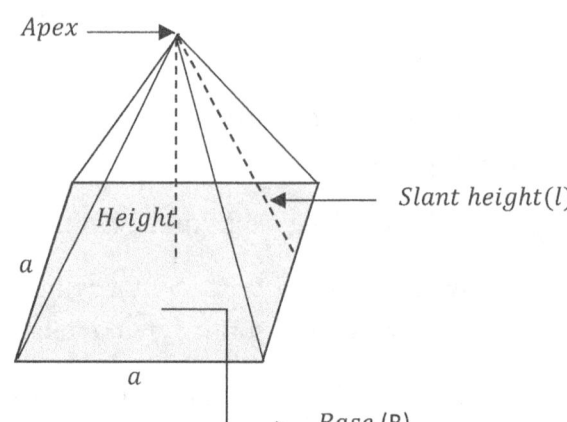

Solution:

- Volume:

$$V = \frac{1}{3} B \cdot h = \frac{1}{3} a^2 \cdot h = \frac{1}{3} \times 6^2 \times 4 = \frac{1}{3} \times 36 \times 4 = 48 \ cm^3.$$

- Total surface area: $S_{total} = a^2 + 2al = 6^2 + 2 \times 6 \times 10 = 36 + 120 = 156 \ cm^2$

Cones

A cone is a three-dimensional geometric shape with a circular base and a single apex. It can be thought of as a pyramid with a circular base.

Key Properties of the cone:

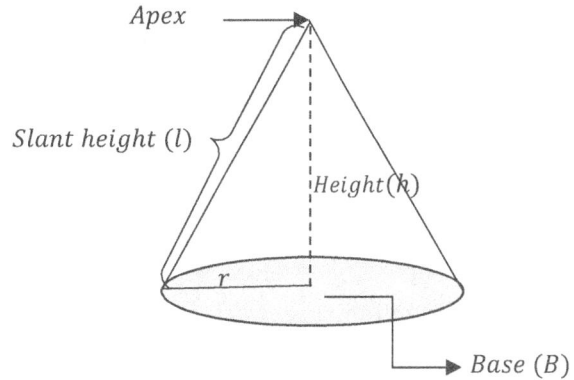

1. **Base**: A circle.
2. **Apex**: The top point where the curved surface meets.
3. **Height (h)**: The perpendicular distance from the base to the apex.
4. **Radius (r)**: The radius of the circular base.
5. **Slant Height (l)**: The distance from the apex to any point on the edge of the base.

Surface Area:

- **Base Area (B)**: $B = \pi r^2$
- **Lateral Surface Area**: $S = \pi r l$
- **Total Surface Area**: $S = \pi r (r + l)$
- ☑ The slant height (l) can be found using the Pythagorean theorem if the height (h) and radius (r) are known. $l = \sqrt{r^2 + h^2}$

Volume: $V = \frac{1}{3}\pi r^2 h$

Example:

Calculate the volume and surface area of a cone with a radius of $3\ in$ and a height of $4\ in$. ($\pi \approx 3$)

Solution:

Calculate the volume: $V = \frac{1}{3}\pi r^2 h = \frac{1}{3} \times 3 \times 3^2 \times 4 = 36\ in^3$

Calculate the slant height: $l = \sqrt{r^2 + h^2} = \sqrt{3^2 + 4^2} = \sqrt{9 + 16} = \sqrt{25} = 5\ in$

Calculate the surface area: $S = \pi r(r + l) = 3 \times 3 \times (3 + 5) = 72\ in^2$

Word Problems

To solve geometry problems and solid figures, you'll need to follow a structured approach.

Step-by-Step Guide:

1. **Read the Problem Carefully**:
   - Understand what the problem is asking.
   - Identify the key information given in the problem.

2. **Draw a Diagram**:
   - If possible, sketch a diagram of the shape or figure involved.
   - Label the given dimensions and other relevant information.

3. **Identify the Relevant Formulas**. It includes all relevant formula about perimeter, area and volume.

4. **Apply the Formulas**:
   - Substitute the given values into the appropriate formulas.
   - Perform the necessary calculations.

5. **Check Your Work**:
   - Verify that your answer makes sense in the context of the problem.
   - Double-check your calculations for accuracy.

Example:

A swimming pool is in the shape of a rectangular prism with dimensions 10 meters in length, 4 meters in width, and 2 meters in height. The pool also has a semicircular hot tub attached to one of the shorter sides, with a radius of 2 meters and the same height as the pool. Calculate the total volume of water the pool and hot tub can hold.

Solution:

1. Identify the Volumes of Each Section:
   $V_{pool} = length \times width \times height = 10 \times 4 \times 2 = 80$ cubic meters
2. Semicircular hot tub:
   $V_{semicircle} = \frac{1}{2} \times \pi r^2 h \approx \frac{1}{2} \times 3.14 \times 2^2 \times 2 = \frac{1}{2} \times 3.14 \times 2 \times 2 \times 2 = 12.56$ cubic meters
3. Add the volume together:
   $V_{total} = V_{pool} + V_{semicircle} = 80 + 12.56 = 92.56$ cubic meters

www.mathnotion.com

Worksheets

⚜ Angles of Triangles and Quadrilateral
Find the value of angle x.

1)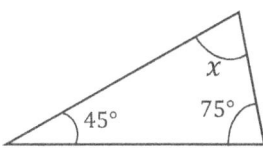

2)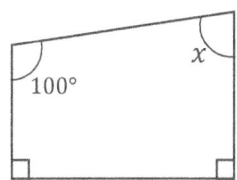

3) The quadrilateral is a square.

   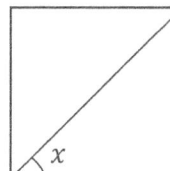

4) The triangle is isosceles, and the dotted line is the angle bisector.

   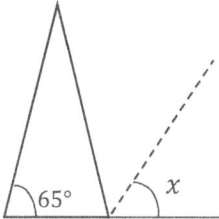

5) The dotted line is the angle bisector.

   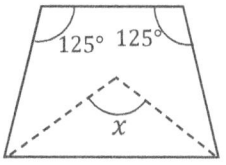

⚜ Interior and Exterior Angles of Polygons
Find the answers.
1) Find the sum of the interior angles of a pentagon.
2) Calculate the measure of each exterior angle of a regular hexagon.
3) If each interior angle of a regular polygon is $144°$, determine the number of sides of the polygon.
4) A regular polygon has 12 sides. Find the measure of each interior angle.
5) The interior angle of a regular polygon is three times the corresponding exterior angle. Determine the number of its sides.

⚜ Angle of Circles
Find the value of angle x.

1)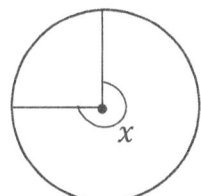

2)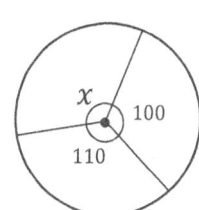

3) 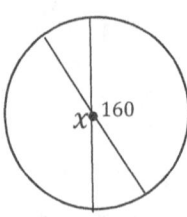   4)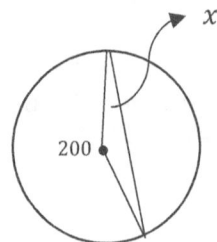

5) The circle is divided into three equal parts from the center.

🔖 Pythagorean Theorem

Find the value of *x* (all units are cm).

1) 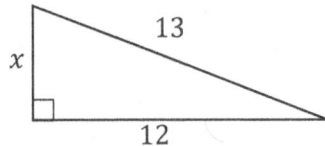   2) 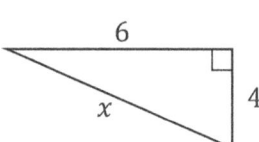   3)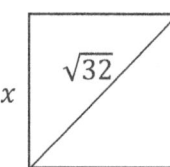

4) 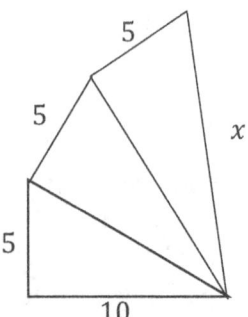   5)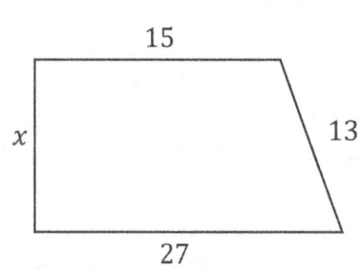

🔖 Area of Compound Figures

Find the area (all units are cm and $\pi \approx 3.14$).

1) 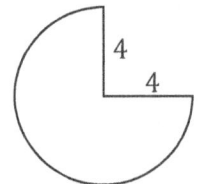   2)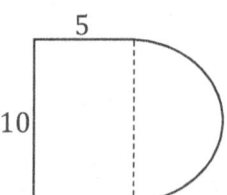

www.mathnotion.com

3)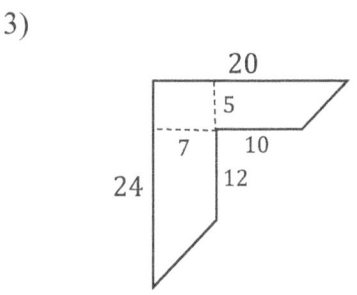

4)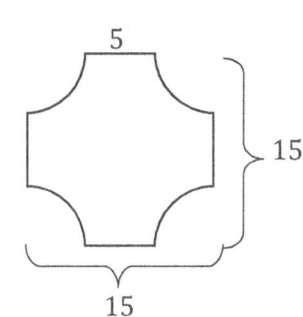

5)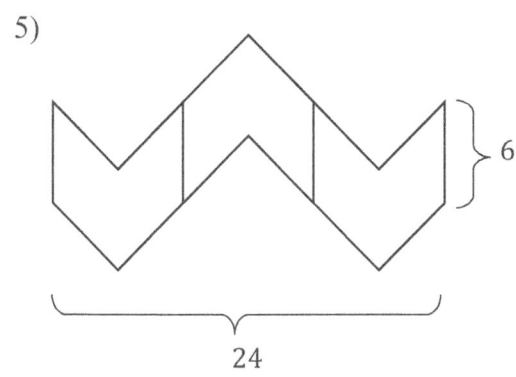

Find the area of gray part, (all units are cm and $\pi \approx 3.14$):

6)

7)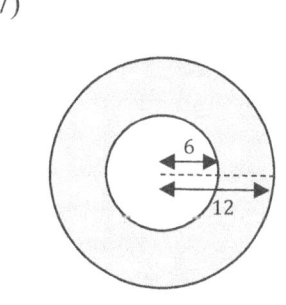

8)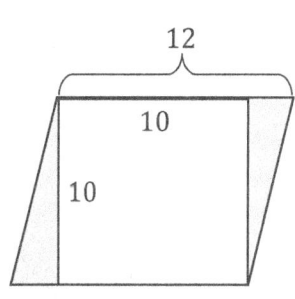

9)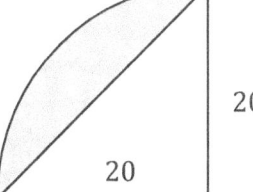

10)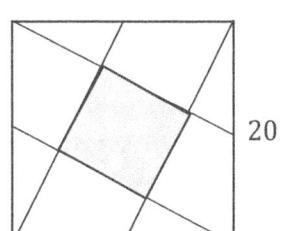

7th Grade Rhode Island Math

🕮 Perimeter of Compound Figures

Find the perimeter of the compound figures (all units are cm and $\pi \approx 3.14$):

1)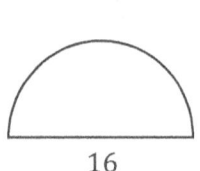

2)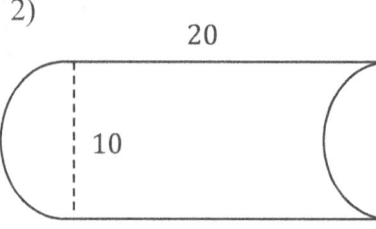

3)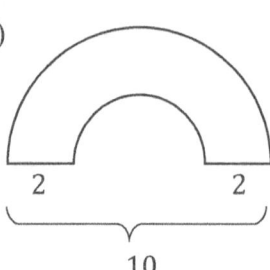

4)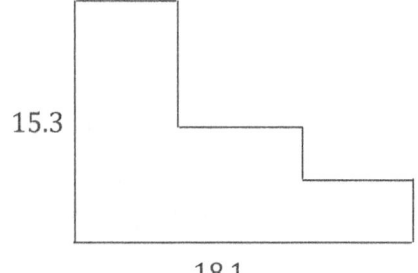

5)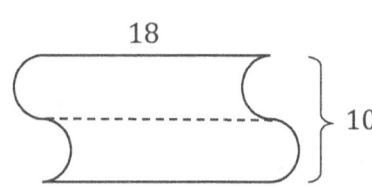

🕮 Front, Side and Top of Three-Dimensional Figures

Determine how each shape is seen from the requested angle of view.

1) Side

2) Top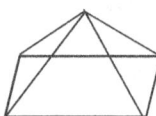

3) Front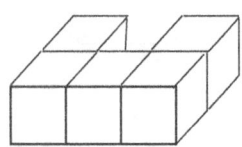

4) Side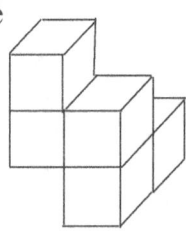

5) Top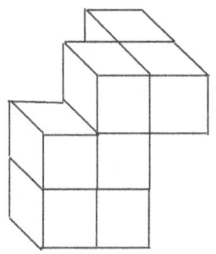

www.mathnotion.com

Cubes

Do the cubes problems.
1) Find the volume and surface area of a cube with a side length of 4.5 ft.
2) Determine the length of the diagonal space of a cube with a side length of 6 cm.
3) If the side length of a cube is doubled, by what factor does the volume of the cube increase?
4) If a cube with a side length of 10 in has a hole with dimensions 3 in by 3 in by 5 in created in it, find the volume of the remaining shape.
5) The total length of all the edges of a cube is 72 m. Find its surface area.

Rectangular Prism

Solve.
1) Find the volume and surface area of a rectangular prism with dimensions 1.2 cm by 4.5 cm by 3.8 cm.
2) A rectangular prism has a volume of 120 cm^3, a width of 6 cm, and a height of 2.4 cm. Find its length.
3) If the length, width, and height of a rectangular prism are each doubled by 2 in, how does the volume change?
4) A rectangular prism is partially filled with water. The prism has dimensions of 10 cm by 8 cm by 12 cm, and the water level is 5 cm high. If more water is added and the water level rises to 10 cm, how much water is added?
5) A rectangular prism with dimensions 8 in by 6 in by 4 in has a smaller rectangular prism with dimensions 2 in by 2 in by 2 in removed from its interior. Find the surface area of the remaining shape.

Cylinder

Calculate ($\pi \approx 3$).
1) Calculate the volume of a cylinder with a radius of 4.9 cm and a height of 10.3 cm.
2) Find the surface area of a cylinder with a radius of 2.8 ft and a height of 6.1 ft.
3) A cylinder has a volume of 200 cubic centimeters and a radius of 5 cm. Find its height.
4) A cylindrical tank with a radius of 3 meters and a height of 10 meters is used to store water. If smaller cylindrical containers with a radius of 1 meter and a height of 2 meters are used to transfer water from the tank, how many containers will be needed to empty the tank?
5) If we pour water from a cylinder with a radius of 10 in and a height of 5 in into a cone with a radius of 8 in and a height of 30 in, to what height will the water rise in the cone?

Pyramids

Find the answers.
1) Find the volume of a square pyramid with a base side length of 4 cm and a height of 6 cm.
2) Calculate the surface area of a square pyramid with a base side length of 5.3 ft and a slant height of 8 ft.
3) A pyramid has a square base with an area of 36 cm^2 and a volume of 108 cm^3. Find the height of the pyramid.
4) A pyramid has a base in the shape of a right triangle with legs of 7.5 in and 8 in. The height of the pyramid is 10 in. Find the volume of the pyramid.

5) Two pyramids are similar, and the ratio of their heights is 2: 3. If the volume of the smaller pyramid is 40 cubic centimeters, find the volume of the larger pyramid.

Cone

Calculate ($\pi \approx 3$).
1) Calculate the volume and surface area of a cone with a radius of 4 mm and a slant height of 7mm.
2) A cone has a volume of 150 cubic centimeters and a radius of 5 cm. Find its height.
3) Determine the lateral surface area of a cone with a radius of 6 meters and a height of 8 meters.
4) If the radius of a cone triples and the height remains the same, how does the volume change?
5) A cone has a radius of 6.4 cm and a height of 0.105 m. A plane cuts the cone horizontally, creating a smaller cone. The height of the smaller cone is 40 mm. Find the volume of the smaller cone using the properties of similar cones.

Word Problems

Solve the following problems:
1) A cylindrical water tank has a radius of 6 meters and a height of 10 meters. The tank is used to store rainwater for the community. How much rainwater (in cubic meters) can the tank hold?
2) A playground has a large cone-shaped structure with a base radius of 7 meters. The maintenance crew needs to cover the base with soft surface material. What is the area of the base that will be covered?
3) Sarah wants to create a decorative planter in her garden using a frustum of a cone. The top radius of the frustum is 6 cm, the bottom radius is 10 cm, and the height is 15 cm. How much soil can the planter hold?
4) A construction worker needs to build a right-angled triangular ramp for a building entrance. The base of the ramp is 6 feet long, and the height is 8 feet. What is the length of the ramp?
5) We have a metal sheet in the shape of a rectangle with dimensions of 0.20 m by 0.12 m. We cut squares with a side length of 0.02 m from each corner. Find the volume of the resulting box.
6) A ladder leans against a wall. The base of the ladder is 2.5 meters away from the wall, and the ladder reaches a height of 6.2 meters on the wall. Find the length of the ladder.
7) We rotate a rectangle with a length of 5 in and a width of 3 in around its width. Find the volume and surface area of the resulting solid.
8) The radius of the base of a cylindrical container is 5 centimeters. If the volume of this cylinder is 1,099 cubic centimeters, what is the maximum height to which water can be filled in this container?
9) We have a swimming pool with dimensions of 10 meters in length, 5 meters in width, and 3 meters in depth. We want to fill this pool using a water tap that delivers 100 liters of water per minute. How many hours will it take to fill the pool?
10) We drop a stone into a conical container with a base radius of 5 centimeters. The water in the container rises by 2 centimeters. What is the volume of the stone?

Answer of Worksheets

Angles of Triangles and Quadrilateral
1) 60°
2) 80°
3) 45°
4) 57.5°
5) 125°

Interior and Exterior Angles of Polygons
1) 540°
2) 60°
3) 10
4) 150°
5) 8

Angle of Circles
1) 270°
2) 150°
3) 160°
4) 10°
5) 30°

13-2. Area and Perimeter

Pythagorean Theorem
1) $5\ cm$
2) $2\sqrt{13}\ cm$
3) $4\ cm$
4) $5\sqrt{7}\ cm$
5) $5\ cm$

Area of Compound Figures
1) $37.68\ cm^2$
2) $89.25\ cm^2$
3) $201\ cm^2$
4) $146.5\ cm^2$
5) $144\ cm^2$
6) $344\ cm^2$
7) $339.12\ cm^2$
8) $20\ cm^2$
9) $114\ cm^2$
10) $100\ cm^2$

Perimeter of Compound Figures
1) $41.12\ cm$
2) $71.4\ cm$
3) $29.12\ cm$
4) $66.8\ cm$
5) $67.4\ cm$

Front, Side and Top of Three-Dimensional Figures

1) Side   2) Top 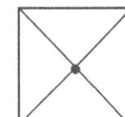 3) Front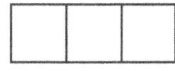

4) Side 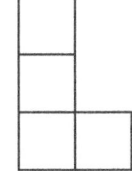 5) Top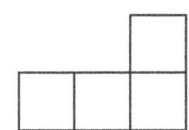

Cubes
1) The volume: 91.125 cm^3, surface area: 121.5 cm^2
2) $6\sqrt{3}$ cm
3) 8
4) 955 in^3
5) 216 square meters

Rectangular Prism
1) The volume: 20.52 cm^3, surface area: 54.12 cm^2
2) The length: $\approx$ 8.33 cm
3) The volume increases by a factor of 8.
4) 400 cm^3
5) 224 in^2

Cylinder
1) $\approx$ 741.9 cm^3
2) 149.52 ft^2
3) $\frac{8}{3} cm$
4) 45 small containers.
5) 23.44 in

Pyramids
1) 32 cm^3
2) 112.89 ft^2
3) 9 cm
4) 100 in^3
5) 135 cm^3

Cone
1) Volume: 91.904 mm^3, Area: 132 mm^2
2) 6 cm
3) 180 m^2
4) The volume of the cone increases by a factor of 9.
5) $\approx$ 23.8 cm^3

Word Problems
1) $\approx$ 1130.4 m^3
2) $\approx$ 153.86 m^2
3) $\approx$ 3077.2 cm^3
4) 10 feet
5) 0.000256 m^3
6) 6.68 meters
7) Volume: $\approx$ 235.5 in^3, surface area: $\approx$ 251.2 in^2
8) $\approx$ 14 cm
9) 25 hours
10) $\approx$ 52.33 cm^3

Chapter 14: Statistics and Probability

Topics that you'll learn in this chapter:

- ✓ Mean and Median
- ✓ Mode and Range
- ✓ Mean Absolute Deviation
- ✓ Frequency Chart
- ✓ Histograms
- ✓ Pie Graph
- ✓ Box and Whisker Plot
- ✓ Quartiles and Outlier
- ✓ Probability of Simple Events
- ✓ Probability of Opposite Events
- ✓ Probability of Mutually Events
- ✓ Experimental Probability
- ✓ Make Predictions
- ✓ Theoretical Probability
- ✓ Compound Events
- ✓ Probability Word Problems
- ✓ Worksheets
- ✓ Answer of Worksheets

Mean and Median

Mean (Average)

Mean is what most people call the "average." To find mean of a set of numbers, you add up all the numbers and then divide by how many numbers there are.

Median

The median is the middle value in a list of numbers when they are arranged in order. If there's an odd number of numbers, the median is the middle one. If there's an even number of numbers, the median is the average of the two middle numbers.

Examples:

1) Consider the following data set:
   23, 25, 26, 28, 34, 35, 36, 37, 45, 46, 47, 48, 56, 57, 58, 59

   Calculate the mean and median of data set.

   Solution:

   - Calculating the Mean:

   $$Mean = \frac{23 + 25 + 26 + 28 + 34 + 35 + 36 + 37 + 45 + 46 + 47 + 48 + 56 + 57 + 58 + 59}{16}$$

   $$= \frac{660}{16} = 41.25$$

   - Calculating the median: Since there are 16 values in the data set (an even number), the median is the average of the 8th and 9th values when the data is arranged in ascending order.

   $$Median = \frac{36+37}{2} = \frac{73}{2} = 36.5$$

2) Find the median of the data:
   85, 92, 78, 88, 95, 80, 82, 91, 94, 87, 84, 89, 76, 90, 81

   Solution:

   1. Arrange the numbers in ascending order:

      76, 78, 80, 81, 82, 84, 85, 87, 88, 89, 90, 91, 92, 94, 95

   2. Determine the middle number: Since there are 15 data points (an odd number), the median is the middle number in the ordered list. The middle number is the 8th number in the list:

      76, 78, 80, 81, 82, 84, 85, **87**, 88, 89, 90, 91, 92, 94, 95

      So, the median of the data set is 87.

Mode and Range

Mode

The mode is the value that appears most frequently in a data set. A data set may have one mode, more than one mode, or no mode at all if all values are unique.

Range

The range is the difference between the maximum and minimum values in a data set. It provides a measure of how spread out the values are.

$$Range = Maximum\ Value - Minimum\ Value$$

Examples:

1) Consider the following data set:

   15, 18, 20, 21, 14, 15, 22, 18, 18, 45, 32, 31, 50, 41, 23, 19, 18

   Calculate the mode and range for this data set.

   Solution:
   - Calculate the mode: To find the mode of a data set, we look for the number that appears most frequently. The number that appears the most frequently is 18, which appears 4 times. Thus, the mode is 18.
   - Calculate the range: $Range = 50 - 14 = 36$

2) The following data set represents the scores of 15 students on a math test. What mode and range of scores? If an additional student scored 78, how would that affect the mode and range?

   88, 72, 88, 93, 85, 78, 79, 88, 91, 78, 85, 93, 94, 77, 94

   Solution:
   - Calculate the mode: Since 88 appears the most frequently (3 times), the mode is: 88
   - Calculate the range: The range is the difference between the highest and lowest values in the data set: $94 - 72 = 22$
   - Add an additional score: If an additional student scored 78, the updated data set is:

     88, 72, 88, 93, 85, 78, 79, 88, 91, 78, 85, 93, 94, 77, 94, 78
   - Mode: both 88 and 78 appear 3 times, making them both modes (bimodal): 88 and 78
   - The range remains unchanged because the highest and lowest values are still 94 and 72: $94 - 72 = 22$

Mean Absolute Deviation

Mean Absolute Deviation (MAD) is a way to measure how spread out the numbers are in a data set. It tells you, on average, how far each number in the data set is from the mean (average) of the data.

Here's how you can calculate the Mean Absolute Deviation (MAD) step by step:

1. **Find the Mean (Average):**

    - Add all the numbers to your data set.
    - Divide that sum by the total number of numbers in the data set.

2. **Find the Differences:**

    - Subtract the meaning from each number in your data set. This gives you the "difference" for each number.
    - Ignore if the result is negative (i.e., take the absolute value of the differences). That's why it's called "absolute deviation."

3. **Find the Average of the Differences:**

    - Add up all the absolute differences.
    - Divide that sum by the total number of numbers in the data set.

The result is the Mean Absolute Deviation (MAD).

Example:

Calculate MAD for the following data set: Data set: 5, 8, 7, 10, 6

Solution:

1. Find the mean: Mean = $(5 + 8 + 7 + 10 + 6) \div 5 = 36 \div 5 = 7.2$

2. Find the difference (and take absolute values)

    - $|5 - 7.2| = 2.2$
    - $|8 - 7.2| = 0.8$
    - $|7 - 7.2| = 0.2$
    - $|10 - 7.2| = 2.8$
    - $|6 - 7.2| = 1.2$

3. Find the mean of the absolute differences:

    Mean absolute deviation: $(2.2 + 0.8 + 0.2 + 2.8 + 1.2) \div 5 = 7.2 \div 5 = 1.44$

    So, the Mean Absolute Deviation (MAD) for this data set is 1.44.

Frequency Chart

Frequency charts are a great way to visualize how often different values or ranges of values occur in a data set.

Types of Frequency Charts:

1. **Frequency Distribution Table**: A table that lists each value (or range of values) and its frequency.
2. **Histogram**: A graphical representation of a frequency distribution, where data is grouped into ranges (bins) and represented by bars.
3. **Bar Chart**: Similar to a histogram but used for categorical data.

Creating a Frequency Distribution Table

Given a data set, we can create a frequency distribution table by counting the occurrences of each value or range of values.

Example:

Create a frequency distribution table for the data set:

27, 30, 31, 32, 40, 41, 43, 46, 48, 50, 50, 51, 52, 59, 62, 66, 68, 73, 75, 77, 82

Solution:

- Determine the Range: Find the difference between the highest and lowest values in the data set. $Range = highest\ value - lowest\ value = 82 - 27 = 55$
- Choose the number of classes: We can decide to have 5 classes or categories.
- Calculate the class width: $Class\ width = \frac{Range}{Number\ of\ Classes} = \frac{55}{5} = 11$
- Set up the classes and create the table: Define the class intervals. Start from the lowest and add the class width to create the intervals. Then present the data in the table format with two columns, one for class intervals and one for the frequencies:

Class interval	Frequency
27 – 38	4
38 – 49	5
49 – 60	5
60 – 71	3
71 – 82	4

www.mathnotion.com

Histograms

Histograms are a powerful tool for visualizing the distribution of numerical data. They provide a graphical representation of data by grouping it into bins or intervals and displaying the frequency of data points in each bin.

How to Create a Histogram

1. **Collect and Sort Data**: Gather your numerical data and sort it in ascending order.

2. **Determine the Number of Bins**: Decide the number of intervals (bins) to divide your data into. The number of bins can affect the readability of your histogram.

3. **Calculate Bin Width**: The bin width is the range of values each bin covers. It can be calculated as: $Bin\ Width = \frac{Max\ Value - Min\ Value}{Number\ of\ Bins}$

4. **Create the Bins**: Divide the data into bins based on the bin width.

5. **Count Frequencies**: Count the number of data points in each bin.

6. **Draw the Histogram**: Plot the bins on the x-axis and the frequencies on the y-axis and draw bars for each bin.

Example:

Create a histogram for following data set in 3 bins.

8, 12, 13, 15, 20, 25, 26, 28, 30, 30, 31, 36, 39, 40, 42, 56

Solution:

1. Calculate bin width: $Bin\ Width = \frac{Max\ Value - Min\ Value}{Number\ of\ Bins} = \frac{56-8}{3} = \frac{48}{3} = 16$

2. Create the bins and count frequencies: In this histogram:
   - The bin for $8 - 24$ contains 5 data points.
   - The bin for $25 - 40$ or $(24 - 40]$ contains 9 data points.
   - The bin for $41 - 56$ or $(40 - 56]$ contains 2 data points.

3. Draw the histogram: The x-axis represents the value ranges (bins). The y-axis represents the frequency of data points within each bin. Each bar shows how many data points fall within each bin range.

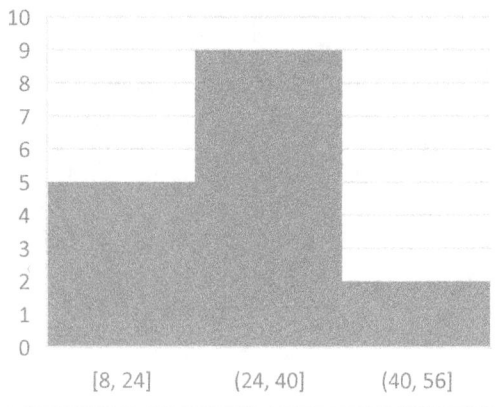

Pie Graph

Pie graphs, also known as pie charts, are a visual representation of data where the whole circle represents the total data, and slices represent parts of the whole. Each slice's size is proportional to the quantity it represents.

Creating a Pie Chart:

1. **Collect Data**: Gather the data you want to represent.
2. **Calculated Percentages**: Determine the percentage that each category represents of the total.
3. **Draw the Pie Chart**: Use a charting tool or software to create the pie chart, where each slice corresponds to a category.

Example:

Consider the following example data set representing the sales distribution of different products:

Product	Sales
Product A	150
Product B	200
Product C	300
Product D	100
Product E	250

Solution:

1. Calculate percentages:
   - Product A: $\frac{150}{1000} = \frac{15}{100} = 15\%$
   - Product B: $\frac{200}{1000} = \frac{20}{100} = 20\%$
   - Product C: $\frac{300}{1000} = \frac{30}{100} = 30\%$
   - Product D: $\frac{100}{1000} = \frac{10}{100} = 10\%$
   - Product A: $\frac{250}{1000} = \frac{20}{100} = 25\%$
2. Draw the circle and label the slices:

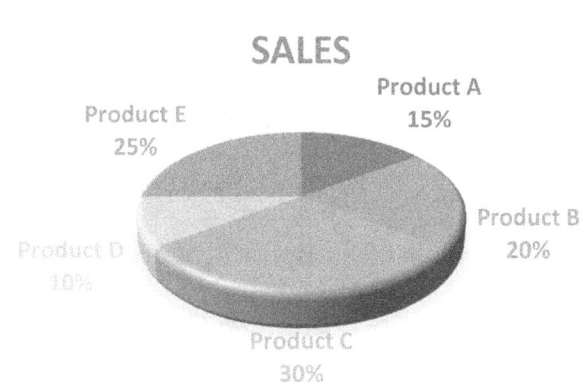

Quartiles and Outlier

Quartiles and outliers are important concepts in statistics that help understand the distribution of a dataset. Quartiles divide a dataset into four equal parts, while outliers are data points that are significantly different from the rest of the data.

Steps to Calculate Quartiles and Identifying Outliers:

1. **Sort the Data**: Arrange the data in ascending order.
2. **Calculate Median (Q2)**
3. **Calculate $Q1$:** The median of the first half of the data.
4. **Calculate $Q3$:** The median of the second half of the data.
5. **Calculate the Interquartile Range (IQR):** $IQR = Q3 - Q1$
6. **Determine the Lower and Upper Boundaries for Outliers:**
   - Lower Boundary: $Q1 - 1.5 \times IQR$
   - Upper Boundary: $Q3 + 1.5 \times IQR$

Any data points below the lower boundary or above upper boundary are considered outliers.

Example:

Calculate quartiles and outliers for monthly savings (in dollars):

150, 200, 180, 220, 170, 3000, 190, 230, 160, 175

Solution:

1. Sort the Data: Arrange the data in ascending order:
   150, 160, 170, 175, 180, 190, 200, 220, 230, 3000
2. Calculate $Q2$: The median of the entire data set: $Q2 = \frac{180+190}{2} = 185$
3. Calculate $Q1$: The median of the first half of the data: $Q1 = 170$
4. Calculate $Q3$: The median of the second half of the data: $Q3 = 220$
5. Calculate the interquartile range (IQR): $IQR = Q3 - Q1 = 220 - 170 = 50$
6. Determine the lower and upper boundaries for outliers:
   - Lower Boundary: $Q1 - 1.5 \times IQR = 170 - 1.5 \times 50 = 170 - 75 = 95$
   - Upper Boundary: $Q3 + 1.5 \times IQR = 220 + 1.5 \times 50 = 220 + 75 = 295$

Any data points below 95 or above 295 are considered outliers. In this data set the score 3,000 is above 295, making it an outlier.

www.mathnotion.com

Box and Whisker Plot

A box and whisker plot, or box plot, is a simple and useful graph that helps us visualize how a set of data is spread out. It shows the minimum, first quartile ($Q1$), median, third quartile ($Q3$), and maximum values, making it easier to see how the data is distributed and identify any outliers.

Steps to Create a Box and Whisker Plot:

1. **Collect Date:** Gather your numerical data.
2. **Calculate Quartiles:** These divide the data set into four equal parts.
   - $Q1$ **(First Quartile)**: The median of the lower half of the data.
   - $Q2$ **(Median)**: The middle value of the entire data set.
   - $Q3$ **(Third Quartile)**: The median of the upper half of the data.
3. **Interquartile Range (IQR)**: The range between $Q1$ and $Q3$. It shows the middle 50% of the data.

$$IQR = Q3 - Q1$$

5. **Whiskers**: The lines that extend from the quartiles to the smallest and largest values within a certain range, usually 1.5 times the IQR.
6. **Outliers**: Any data points that fall outside the range of the whiskers are considered outliers.

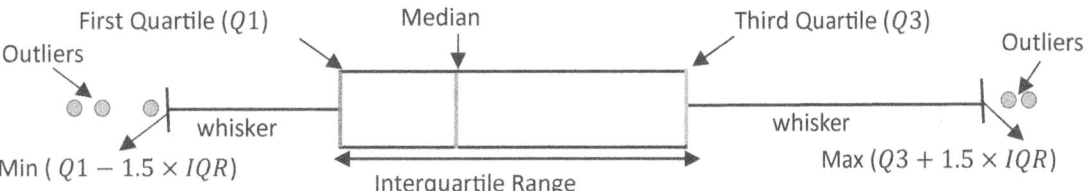

Example:

Consider the following data set and create a box and whisker plot:

60, 65, 90, 95, 110, 125, 129, 130, 130, 145, 155

Solution:

1. Calculate quartiles: $Q2 = 125, Q1 = 90$ and $Q3 = 130$
2. Interquartile range (IQR): $IQR = Q3 - Q1 = 130 - 90 = 40$
3. Minimum or lower boundary: $Q1 - 1.5 \times IQR = 90 - 1.5 \times 40 = 30$
4. Maximum or upper boundary: $Q3 + 1.5 \times IQR = 130 + 1.5 \times 40 = 190$
5. Outliers: There is no outlier in this data set.

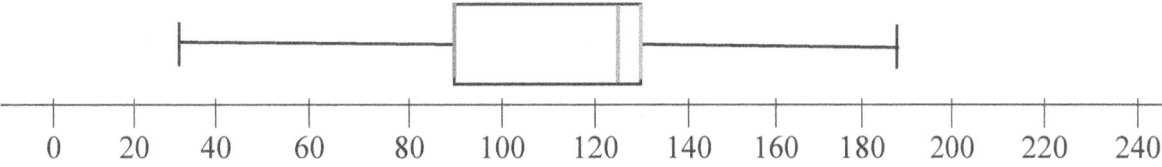

www.mathnotion.com

Probability of Simple Events

The probability of a simple event is a measure of the likelihood that the event will occur. It is calculated by dividing the number of favorable outcomes by the total number of possible outcomes. The probability of an event A is denoted by $P(A)$.

Basic Probability Formula

$$P(A) = \frac{\text{Number of favorable outcomes}}{\text{Total number of possible outcoms}}$$

Key Concepts

1. **Sample Space** (S): The set of all possible outcomes.

2. **Event** (A): A subset of the sample space, consisting of outcomes that satisfy a certain condition.

3. **Favorable Outcomes**: The outcomes in the event A.

Examples:

1) Consider drawing a single card from a standard 52-card deck. Calculate the possibility of drawing an Ace.
   Solution:
   - Total number of possible outcomes: $S = \{all\ 52\ cards\}$
   - Favorable outcomes: Ace of Spades, Ace of Hearts, Ace of Diamonds, Ace of Clubs}
   - Number of favorable outcomes: 4.
   - The probability of drawing an Ace can be calculated using the formula:

   $$P(drawing\ an\ Ace) = \frac{4}{52} = \frac{1}{13}$$

2) A jar contains 10 red marbles, 15 blue marbles, and 5 green marbles. If you pick one marble at random from the jar, what is the probability that it will be blue marble?

   Solution:
   - Total number of possible outcomes: There are 10 red marbles + 15 blue marbles + 5 green marbles = 30 marbles in total.
   - Number of favorable outcomes: There are 15 blue marbles.
   - The probability P of picking a blue marble can be calculated using the formula:

   $$p\ (blue\ marble) = \frac{15}{30} = \frac{1}{2}$$

Probability of Opposite Events

The probability of opposite events, also known as complementary events, is a key concept in probability. Complementary events are two outcomes that are the only possible outcomes and cannot occur at the same time. For example, if you're rolling a fair six-sided die, the complementary events are "rolling a 6" and "not rolling a 6."

Understanding Complementary Events:

1. **Definition**: If an event A happens, the complement of A (denoted as A' or $\bar{A}$) is the event that A does not happen.

2. **Probability Rule**: The sum of the probabilities of an event and its complement is always 1.

$$P(A) + P(A') = 1$$

3. **Calculating the Complement**: If you know the probability of event A, you can find the probability of its complement by subtracting the probability of A from 1.

$$P(A') = 1 - P(A)$$

Examples:

1) You roll a standard six-sided die. What is the probability that you do not roll a 4?
   Solution:
   1. Total number of possible outcomes: There are 6 faces on a die, so there are 6 possible outcomes (1, 2, 3, 4, 5, 6).
   2. Number of favorable outcomes (rolling a 4): There is only 1 face with a 4 on it.
   3. The probability of rolling a 4 is:

   $$P(Rolling\ a\ 4) = \frac{1}{6}$$

   4. The probability of **not** rolling a 4, which is the complement of rolling a 4, can be calculated as:

   $$P(Not\ rolling\ a\ 4) = 1 - P(Rolling\ a\ 4) = 1 - \frac{1}{6} = \frac{5}{6}$$

2) You roll a standard six-sided die. What is the probability that you will **not** roll a number less than 3?
   Solution:
   1. Total number of possible outcomes: There are 6 faces on a die, so there are 6 possible outcomes (1, 2, 3, 4, 5, 6).
   2. Number of favorable outcomes (rolling a 4): The numbers less than 3 are 1 and 2.
   3. The probability of rolling a number less than 3 is: $P(number\ less\ than\ 3) = \frac{2}{6} = \frac{1}{3}$
   4. The probability of **not** rolling a number less than 3 is:

   $$P(Not\ rolling\ a\ number\ less\ than\ 3) = 1 - P(number\ less\ than\ 3) = 1 - \frac{1}{3} = \frac{2}{3}$$

www.mathnotion.com

Probability of Mutually Events

In probability theory, events can either be mutually exclusive (disjoint) or overlapping (non-disjoint).

Mutually Exclusive Events:

Mutually exclusive events are events that cannot happen at the same time. If one event occurs, the other cannot. The probability of either of two mutually exclusive events occurring is the sum of their individual probabilities.

Key Concept: For mutually exclusive events A and B:

$$P(A \text{ and } B) = 0$$

$$P(A \text{ or } B) = P(A) + P(B)$$

Overlapping Events:

Overlapping events (or non-mutually exclusive events) are events that can happen at the same time. The probability of both overlapping events occurring is the sum of their individual probabilities minus the probability of both occurring together.

Key Concept: For overlapping events A and B:

$$P(A \text{ and } B) \neq 0$$

$$P(A \text{ or } B) = P(A) + P(B) - P(A \text{ and } B)$$

Examples:

1) In a deck of 52 cards, what is the probability of drawing either a Queen or a King?
   Solution:
   Since drawing a Queen and drawing a King are mutually exclusive events (you can't draw both a Queen and a King in a single draw), we can simply add their probabilities.
   1. Probability of drawing a Queen ($P(Q)$): There are 4 Queens in a deck of 52 cards. $P(Q) = \frac{4}{52} = \frac{1}{13}$
   2. Probability of drawing a King ($P(K)$): There are 4 Kings in a deck of 52 cards. $P(K) = \frac{4}{52} = \frac{1}{13}$
   3. Since the events are mutually exclusive: $P(Q \text{ or } K) = P(Q) + P(K) = \frac{1}{13} + \frac{1}{13} = \frac{2}{13}$

2) In a survey of 100 people, 60 like coffee, 40 like tea, and 25 like both coffee and tea. What is the probability of a person liking either coffee or tea?
   Solution:
   Here, liking coffee and liking tea are overlapping events since some people like both.
   1. Probability of liking coffee ($P(C)$): $P(C) = \frac{60}{100} = \frac{6}{10} = 0.6$
   2. Probability of liking tea ($P(T)$): $P(T) = \frac{40}{100} = \frac{4}{10} = 0.4$
   3. Probability of liking both coffee and tea ($P(C \text{ and } T)$): $P(C \text{ and } T) = \frac{25}{100} = 0.25$
   4. Probability of a person liking either coffee or tea:
      $P(C \text{ or } T) = P(C) + P(T) - P(C \text{ and } T) = 0.6 + 0.4 - 0.25 = 0.75$

Theoretical Probability

Theoretical probability is the likelihood of an event occurring based on all possible outcomes, assuming that each outcome is equally likely. Actually, theoretical probability relies on mathematical reasoning and known information about the situation.

Formula:

$$P(A) = \frac{Number\ of\ favorable\ outcomes}{Total\ number\ of\ possible\ outcomes}$$

Steps to Calculate Theoretical Probability:

1. **Identify the total number of possible outcomes:** Determine the total number of outcomes in the sample space.

2. **Identify the number of favorable outcomes:** Determine the number of outcomes that correspond to the event of interest.

3. **Use the probability formula:** Plug the numbers into the formula to find the probability.

Example:

A box contains 4 yellow balls, 3 blue balls, and 5 purple balls. If a ball is randomly selected from the box, what is the probability that the ball is either yellow or purple?

Solution:

Step 1: Determine the Number of Favorable Outcomes:

1. Number of yellow balls: 4

2. Number of purple balls: 5

Since we want the probability of selecting either a yellow or purple ball, we add the number of yellow and purple balls together: Number of favorable outcomes = 4 + 5 = 9

Step 2: Determine the Total Number of Possible Outcomes:

1. Total number of balls: 4 + 3 + 5 = 12

Step 3: Calculate the Theoretical Probability The theoretical probability of selecting either a yellow or purple ball is:

$P(Yellow\ or\ Purple) = \frac{Number\ of\ favorable\ outcomes}{Total\ number\ of\ possible\ outcomes} = \frac{9}{12} = \frac{3}{4}$

So, the probability of selecting either a yellow or purple ball is $\frac{3}{4}$

Experimental Probability

Experimental probability is the probability of an event based on actual experiments and observations. Unlike theoretical probability, which is based on the possible outcomes in a perfect scenario, it provides a realistic measure of probability that can be applied to various fields such as quality control, sports, weather forecasting, and medical research.

Formula for Experimental Probability

The experimental probability of an event A is given by:

$$P(A) = \frac{Number\ of\ times\ event\ A\ occures}{Total\ number\ of\ trials}$$

Steps for Calculating Experimental Probability

1. **Conduct the experiment**: Perform the trials and observe the outcomes.
2. **Count the occurrences**: Record the number of times the event of interest occurs.
3. **Calculate the total number of trials**: Sum up the total number of trials conducted.
4. **Use the formula**: Plug the numbers into the formula for experimental probability.

Example:

Suppose you have a bag with red, blue, and green balls. You draw a ball, record its color, and then put it back in the bag. You repeat this process 100 times and get the following results:

- Number of times you draw a red ball: 45
- The number of times you draw a blue ball: 35
- Number of times you draw a green ball: 20

Calculate the experimental probability of drawing a blue ball.

Solution:

1. Number of times event A (drawing a blue ball) occurs: 35
2. Total number of trials: 100

$$P(drawing\ a\ blue\ ball) = \frac{35}{100} = 0.35$$

So, the experimental probability of drawing a blue ball is 0.35.

Make Predictions

Experimental probability helps us predict future outcomes based on past data from experiments or observations. By calculating experimental probability and applying it to the desired number of future trials, we can estimate the likelihood and frequency of various outcomes.

Steps to Make Predictions:

1. **Conduct the experiment or gather data.**

2. **Calculate experimental probability** using the formula:

$$P(A) = \frac{Number\ of\ times\ event\ A\ occures}{Total\ number\ of\ trials}$$

3. **Use the experimental probability to predict future outcomes** by multiplying the probability by the number of future trials

Examples:

1. Suppose a basketball player makes 75 out of 100 free throw attempts in practice. Based on this experimental probability, predict how many free throws the player will make in the next 40 attempts:

Solution:

1. Calculate the experimental probability of making a free throw:

    - Number of successful free throws: 75
    - Total number of attempts: 100
    - The experimental probability of making a free throw:
    $P(making\ a\ free\ throw) = \frac{75}{100} = 0.75$

2. Make a prediction:

Using experimental probability, we can predict the number of successful free throws in the next 40 attempts:

$P(Predicted\ successful\ free\ throws) = P(making\ a\ free\ throw) \times Number\ of\ future\ attempts = 0.75 \times 40 = 30$

The player is likely to make 30 free throws out of the next 40 attempts based on the experimental probability.

Compound Events

Compound events involve the combination of two or more simple events.

Main types of compound events:
1. **Independent Events**
2. **Dependent Events**

Independent Events:

Definition: Two events are independent if the occurrence of one event does not affect the occurrence of the other. The probability of both independent events A and B occurring is:

$$P(A \text{ and } B) = P(A) \times P(B)$$

Dependent Events:

Definition: Two events are dependent on if the occurrence of one event affects the occurrence of the other. The probability of both dependent events A and B occurring is:

$$P(A \text{ and } B) = P(A) \times P(B \mid A)$$

where $P(B \mid A)$ is the probability of B occurring given that A has already occurred.

Examples:

1) Consider rolling a die and flipping a coin. Find the probability of rolling a 4 and getting heads.
   Solution:
   1. Probability of rolling a 4 (Event A): $P(A) = \frac{1}{6}$
   2. Probability of getting heads (Event B): $P(B) = \frac{1}{2}$

   Since rolling a die and flipping a coin are independent events:
   $$P(A \text{ and } B) = P(A) \times P(B) = \frac{1}{6} \times \frac{1}{2} = \frac{1}{12}$$

2) Suppose we have a deck of 52 playing cards. Find the probability of drawing an Ace and then drawing a King without replacement.
   Solution:
   1. Probability of drawing an Ace (Event A): $P(A) = \frac{4}{52} = \frac{1}{13}$
   2. Probability of drawing a King after an Ace is drawn (Event $B|A$): After drawing an Ace, there are 51 cards left, and still 4 Kings: $P(B \mid A) = \frac{4}{51}$

   Since drawing cards without replacement makes these events dependent:
   $$P(A \text{ and } B) = P(A) \times P(B \mid A) = \frac{1}{13} \times \frac{4}{51} = \frac{4}{663}$$

Probability Word Problems

Let's solve some probability word problems to see how we can apply the concepts of independent and dependent events in real-life scenarios.

Problem 1: Combination of Independent and Dependent Events

A jar contains 6 white and 4 black balls. You pick a ball, do not replace it, and pick another ball. What is the probability that both balls are black?

Solution:

1. **Identify the probability of each event:**

   - Probability of picking a black ball first (Event A): $P(A) = \frac{4}{10} = \frac{2}{5}$

   - Probability of picking a black ball second given the first was black (Event B|A):

   $P(B \mid A) = \frac{3}{9} = \frac{1}{3}$

2. **Since the ball is not replaced, the events are dependent.**

$$P(A \text{ and } B) = P(A) \times P(B \mid A) = \frac{2}{5} \times \frac{1}{3} = \frac{2}{15}$$

Therefore, the probability of picking two black balls in a row is $\frac{2}{15}$.

Problem 2: Probability in a Game of Dice

You roll two six-sided dice. What is the probability of getting a sum of 7?

Solution:

1. **Identify the favorable outcomes:**

   - Possible pairs: $(1,6), (2,5), (3,4), (4,3), (5,2), (6,1)$

2. **Total number of possible outcomes:** $6 \times 6 = 36$

3. **Number of favorable outcomes:** 6

4. **Calculate the probability:**

$$P(\text{sum of } 7) = \frac{6}{36} = \frac{1}{6}$$

Therefore, the probability of getting a sum of 7 when rolling two six-sided dice is $\frac{1}{6}$

Worksheets

🕮 Mean and Median

Find the mean of following data set:
1) $18, 25, 13, 45$
2) $1.2, 0.56, 2.6, 8.4$
3) $87, 88, 89, 90, 91, 92, 93$

Find the median of following data set:
6) $21, 4, 0, 18, 23, 0, 12$
7) $4.1, 4.03, 4, 5.2, 4.8, 3.19$
8) $7.19, -6, 8.45, -2.7, 8.97, 1$

4) $1\frac{2}{3}, 5\frac{3}{4}, 1\frac{1}{6}$
5) $-2.5, 0, 8.61, -18.3, 5.4$

9) $2\frac{1}{5}, 1\frac{7}{8}, \frac{12}{5}, \frac{1}{4}$
10) $7\frac{1}{2}, -2.3, -\frac{1}{5}, 7.49, -\frac{14}{2}$

🕮 Mode and Range

Find the mode of following data set:
1) $7, 2, 0, 2, 11, 2, 7, 17, 12, 2, 21, 0$
2) $8.6, 9, -8.6, 1.5, -9, 1.5, 8.9, 9, 9$
3) $1\frac{2}{5}, \frac{14}{15}, 1\frac{4}{10}, \frac{7}{5}, \frac{16}{10}$

Find the range of following data set:
6) $142, 287, 502, 116, 178$
7) $45, -12, 0, 84, -3, -99$
8) $6.78, 1.05, 7.34, 9, 4.36, 7.14$

4) $\frac{3}{4}, 2.4, 1.75, 2\frac{2}{5}, 0.75, \frac{24}{10}$
5) $-8.3, -\frac{15}{7}, 8.3, -2\frac{2}{14}, 8\frac{3}{11}$

9) $-2\frac{4}{8}, -4\frac{6}{10}, -7\frac{1}{4}, 1, \frac{9}{10}, 3\frac{4}{5}, -3$
10) $11.2, -\frac{1}{3}, 15.1, \frac{8}{15}, -29, 1.35, \frac{9}{20}$

🕮 Mean Absolute Deviation

Find the MAD of the data set:
1) $18, 15, 26, 75, 12$
2) $2, -8, 9, -12, 36, 28, -6$
3) $1.3, 8, 4.6, 9.2, 4.14, 24.4$

4) $12, 36, 12.4, \frac{1}{4}, \frac{3}{8}, 0.15, 4.2$
5) $-14, 2, 0, 3.5, -4, 5.12, -1$

🕮 Frequency Chart

Create a frequency distribution table.
1) $12, 15, 20, 21, 23, 28, 31, 37, 39, 40, 47, 48$ (4 classes)
2) $37, 45, 52, 14, 26, 27, 33, 36, 45, 49, 53, 55$ (3 classes)
3) $14, 25, 1, 36, 12, 8, 22, 32, 10, 15, 18, 20, 30$ (5 classes)
4) $-1, 6, -14, 18, -10, -8, -2, 11, 2, 16, 0, 5$ (4 classes)

🕮 Histograms

Create a histogram.
1) $12, 15, 18, 18, 19, 24, 25, 25, 25, 27, 30$ (group them into 3 bins)
2) $0, 1, 1, 2, 3, 3, 5, 6, 7, 7, 8, 9, 9, 10, 11, 12, 12, 13, 15, 16, 20, 21, 22, 25, 26, 27, 30, 35$ (group them into 4 bins)
3) $17, 21, 23, 23, 37, 39, 39, 41, 42, 43, 43, 47, 49, 50, 51, 52, 53$ (group them into 3 bins)

www.mathnotion.com

4) 6.5, 6.8, 7, 7.2, 8, 8, 8.7, 9.1, 10, 10.5, 12, 12.5, 13, 13.6, 14.2, 14.5, 15 (group them into 5 bins)

Answer question 5 – 10 according to the histogram below:

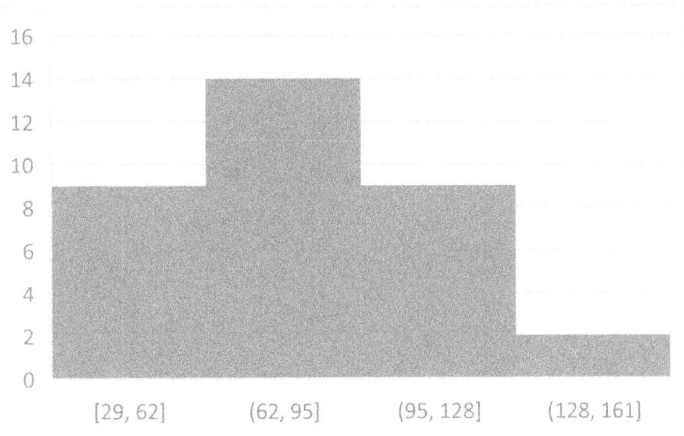

5) Which bin has the lowest frequency of data?
6) Which bin has the highest frequency of data?
7) Which bins have the same number of data?
8) How many numbers are there in this dataset?
9) How many data are there between 62 to 128?
10) Estimate the mean of total data?

Pie Graph

Draw a pie chart according to the information in the following tables:

1) Favorite ice cream flavors:

Vanilla	20 students
Chocolate	25 students
Strawberry	15 students
Mint	10 students
Other	10 students

2) Customer's preference for buying smartphone:

Brand A	200 customers
Brand B	150 customers
Brand C	100 customers
Brand D	30 customers
Brand E	20 customers

3) Distribution of annual budget:

Marketing	$300,000
Research and Development	$250,000
Operation	$200,000
Human Resources	$150,000
IT	$100,000

4) Preferred programming languages among the employees of a company:

Python	45%
JavaScript	25%
Java	15%
C#	10%
Other	5%

Answer question 5 – 10 according to following pie chart:

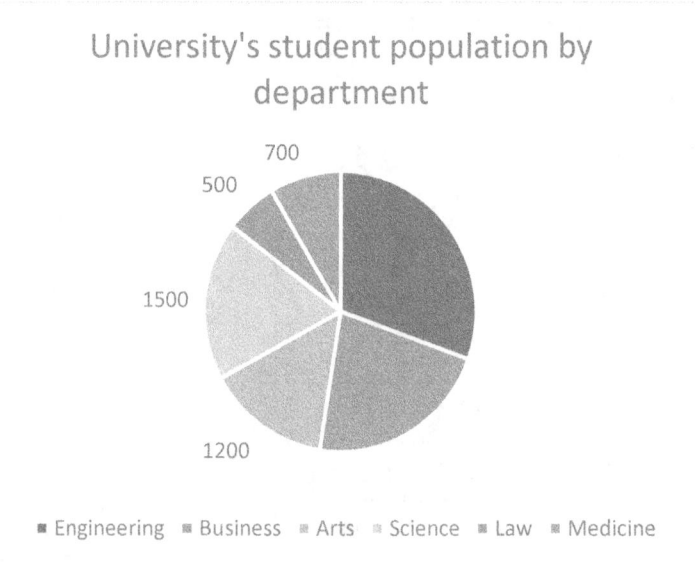

5) What percentage of total students, study in the field of medicine?

6) What percentage of total students, study in the field of law and science?

7) If 15% of students study in the field of art, what is the total number of students in university?

8) If 30% of students study in the field of engineering, how many students study in the field of engineering?

9) What percentage of student's study business?

10) What is the ratio of art students to medical students?

❧ Quartiles and Outlier

Calculate the quartiles and outliers (if there are outliers).
1) 25, 17, 49, 25, 19, 7, 78, 65
2) 125, 115, 100, 130, 145, 170, 195, 200, 110, 100, 95
3) 14, −5, −10, 20, 15, 12, −1, 0, 12, 9
4) 2.6, 3.11, 0.2, 4.35, 59, 2.4, 7.1, 1.6, 8, 2.3
5) 17, 10, 29, 7.8, 2.65, −4, −22, 87, 29

❧ Box and Whisker Plot

Create the box and whisker plot.
1) 10, 11, 11, 12, 18, 20, 21
2) 78, 85, 92, 88, 76, 94, 91, 87, 82, 79
3) 190, 185, 200, 175, 188, 195, 180, 192, 205, 178, 198, 183
4) 1.5, 2, 2.1, 2.5, 3.1, 3.2, 3.5, 3.8, 3.8, 4
5) −7, −5, −2, 0, 6, 7, 8, 8, 10, 12

❧ Probability of Simple Events

Solve the problems below.
1) A bag contains 3 red marbles and 5 blue marbles. What is the probability of randomly picking red marble from the bag?
2) A spinner is divided into 4 equal sections: red, blue, green, and yellow. If the spinner is spun once, what is the probability of landing on either red or yellow?
3) A standard deck of cards is shuffled. What is the probability of drawing a face card (jack, queen, or king)?
4) In a room, there are 30 people, each born in a different year. What is the probability that a randomly chosen person was born in a year that is a leap year (a year divisible by 4)?
5) A club has 100 members, and each member's birthday falls in one of the 12 months of the year. The club leader wants to randomly pick a person who was born in a month with 31 days. What is the probability of selecting such a person?

❧ Probability of Opposite Events

Find answers.
1) A die is rolled once. What is the probability of not rolling a 6?
2) A spinner has 10 equal sections, numbered from 1 to 10. What is the probability of not landing on a number greater than 7?
3) A deck of cards contains 52 cards. If one card is drawn at random, what is the probability that the card is **not** a heart or a diamond?
4) A bag contains 5 red marbles, 8 green marbles, and 12 blue marbles. Two marbles are drawn randomly without replacement. What is the probability that neither of the marbles drawn is green?
5) A family has 4 children. What is the probability that **none** of the children is a girl?

🕮 Probability of Mutually Events

Find the probabilities.
1) You roll a fair die. What is the probability of rolling a 3 or a 5?
2) In a deck of cards, there are 52 cards. What is the probability of drawing a face card or a red card?
3) A family has 3 children. What is the probability that the children are either all boys or all girls?
4) In a city, the weather forecast for tomorrow predicts a 70% chance of rain and a 30% chance of no rain. What is the probability that it will either rain or not rain tomorrow?
5) A theater shows movies in three categories: Action, Comedy, and Drama. Of the 150 movies shown:
   - 60 are Action movies
   - 50 are Comedy movies
   - 40 are Drama movies
   - 30 are both Action and Comedy movies
   - 20 are both Comedy and Drama movies
   - 10 are all three categories: Action, Comedy, and Drama

   What is the probability that a randomly chosen movie is either an Action movie or a Comedy movie?

🕮 Theoretical Probability

Solve probabilities.
1) A fair die is rolled once. What is the probability of rolling a number dividend by 3?
2) Two fair dices are rolled. What is the probability that the sum of the numbers on the dice is greater than 8?
3) In a group of 5 people, what is the probability that at least two people share the same birthday month?
4) A factory produces light bulbs, and 5% of the bulbs are defective. If a box contains 8 bulbs, what is the probability that exactly 2 of them are defective?
5) A clothing store sells 5 red, 7 blue, and 3 green T-shirts. A customer randomly selects 3 T-shirts without replacement. What is the probability that the 3 T-shirts are blue?

🕮 Experimental Probability

Find the experimental probability.
1) You flip a coin 10 times and get 6 heads and 4 tails. What is the experimental probability of getting heads based on these results?
2) A basketball player attempts 20 free throws. The player makes 12 of them. What is the experimental probability that the player makes a free throw?

3) A weather app predicts rain 5 times in a week, and it rains 4 out of those 5 times. What is the experimental probability that it will rain based on the app's predictions?

4) A thermometer records the high temperature for 10 days. The temperatures are: 60, 62, 58, 60, 65, 62, 63, 64, 61, *and* 59 degrees. What is the experimental probability that the temperature is 60 degrees or higher?

5) You toss a penny 50 times and record the results: 30 heads and 20 tails. You then toss the penny 10 more times, and it lands on heads 8 times. What is the experimental probability of getting heads in the next 10 tosses, based on all 60 tosses?

Make Predictions

Find the answers.

1) You flip a coin 10 times, and the result is 6 heads and 4 tails. Based on this data, what is the experimental probability of getting heads? If you flip the coin 20 more times, how many heads do you predict?

2) You roll a fair 6-sided die 12 times, and the results are: 2, 4, 6, 3, 5, 2, 3, 4, 6, 2, 5, 1. Based on this data, what is the experimental probability of rolling a 2? If you roll the die 30 more times, how many times do you predict you will roll a 2?

3) A soccer team plays 40 games, winning 24 and losing 16. Based on this data, what is the experimental probability of the team winning a game? If the team plays 20 more games, how many wins do you predict they will have?

4) You roll two fair dice 50 times, and the sum of the dice is greater than 7 in 15 of those rolls. Based on this data, what is the experimental probability of the sum being greater than 7? If you roll the dice 100 more times, how many times do you predict the sum will be greater than 8?

5) A city has a population of 100,000 people. A survey is conducted where 500 residents are randomly selected. The survey finds that 150 of the respondents own pets. Based on this data, what is the experimental probability that a person in this city owns a pet?
If a random sample of 2,000 people is selected, how many people would you predict own pets based on the experimental probability?

Compound Events

Find the solution to each questions.

1) A standard deck of 52 playing cards is shuffled. You draw two cards without replacement. What is the probability that both cards are queens?

2) You roll two fair 6-sided dice. What is the probability that the sum of the numbers on the dice is 7 and that one die shows an even number?

3) The forecast predicts a 60% chance of rain and a 30% chance of snow for tomorrow. What is the probability that it will rain and snow (assuming the two events are independent)?

4) A basketball player has a 70% chance of making a free throw. If the player attempts two free throws, what is the probability that the player makes both free throws?

www.mathnotion.com

5) A bag contains 4 red, 6 blue, and 10 green marbles. You draw two marbles without replacement. What is the probability that the first marble is red, and the second marble is blue?

Probability Word Problems

Answer the questions below.

1) You flip a fair coin twice. What is the probability of getting heads on both flips?
2) The forecast predicts a 66% chance of rain tomorrow. What is the probability that it will **not** rain tomorrow?
3) In a class of 25 students, 16 vote for pizza and 9 vote for burgers. Based on this data, what is the experimental probability that a randomly selected student will vote for pizza?
4) A clothing store sells 5 red, 7 blue, and 3 green T-shirts. If a customer buys one T-shirt, what is the probability that the T-shirt is either red or blue?
5) You roll two fair 6-sided dice. What is the probability that the sum of the dice is 9 and that one of the dice shows a number greater than 4?
6) You toss a fair coin and roll a fair 6-sided die. What is the probability that the coin lands on heads and the die shows an odd number?
7) A teacher conducts an experiment where she randomly selects 50 students from a class of 100 and asks them about their favorite fruit. The results show that 30 prefer apples and 20 prefer bananas. Based on the experimental data, what is the probability that a randomly chosen student prefers apples?
8) We flip three coins simultaneously. Calculate the probability that at least two of the coins show heads.
9) In a bag, there are three types of marbles: white, red, and yellow. If there are 5 white marbles and 4 red marbles in the bag, and the probability of drawing yellow marble is $\frac{1}{4}$, find the number of yellow marbles in the bag.
10) By throwing a dart, we can score 2, 3, or 6 points if the dart hits the target. If the dart misses the target, the score will be zero. What is the probability that the score will be a multiple of 3?

Answer of Worksheets

Mean and Median
1) 25.25
2) 3.19
3) 90
4) $2\frac{31}{36}$
5) -1.358
6) 12
7) 4.065
8) 4.095
9) $2\frac{3}{80}$
10) $-\frac{1}{5}$

Mode and Range
1) 2
2) 9
3) $\frac{7}{5}$
4) 2.4
5) $-\frac{15}{7}$
6) 386
7) 183
8) 7.95
9) 11.05
10) 44.1

Mean Absolute Deviation
1) 18.32
2) ≈ 14.86
3) ≈ 5.46
4) ≈ 9.25
5) ≈ 4.45

Frequency Chart

1)

Class interval	Frequency
12 – 21	3
21 – 30	3
30 – 39	2
39 – 48	4

2)

Class interval	Frequency
14 – 28	3
28 – 42	3
42 – 56	6

3)

Class interval	Frequency
1 – 8	1
8 – 15	4
15 – 22	3
22 – 29	2
29 – 36	3

4)

Class interval	Frequency
−14 – −6	3
−6 – 2	3
2 – 10	3
10 – 18	3

Histograms

1)
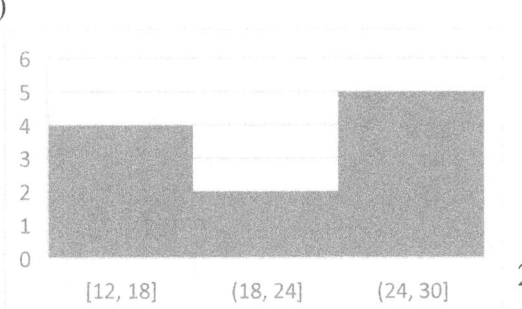

2)
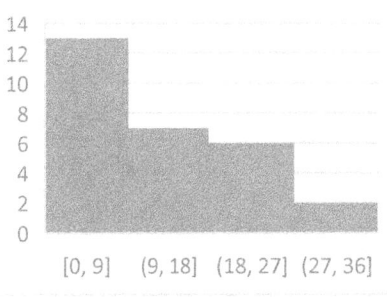

3)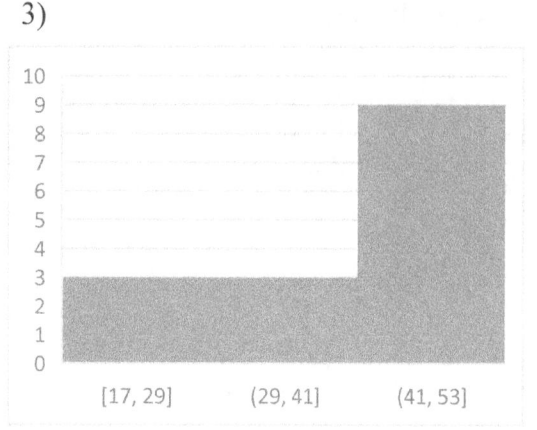

4)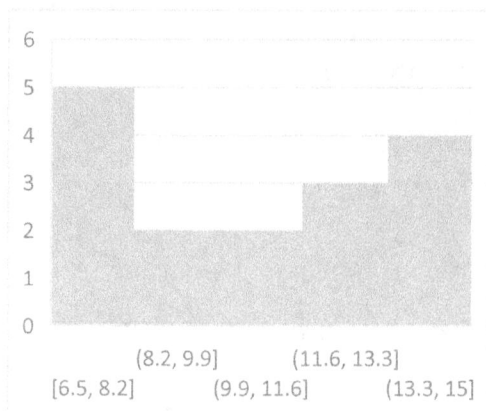

5) 128-161
6) 62-95
7) First and third bins

8) 34
9) 23
10) ≈ 82.3

Pie Graph

1)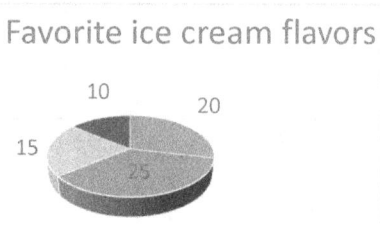

2)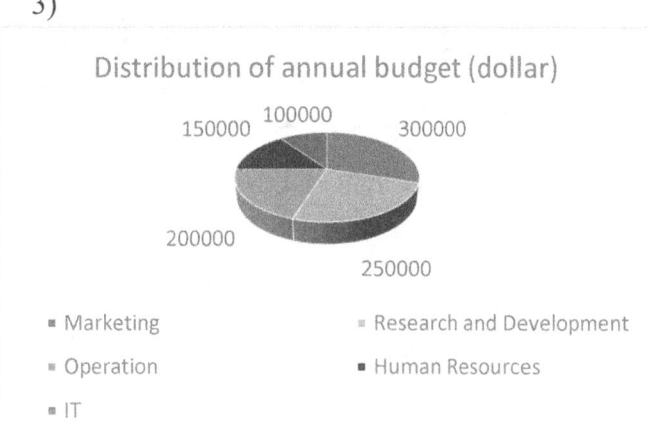

3)

4)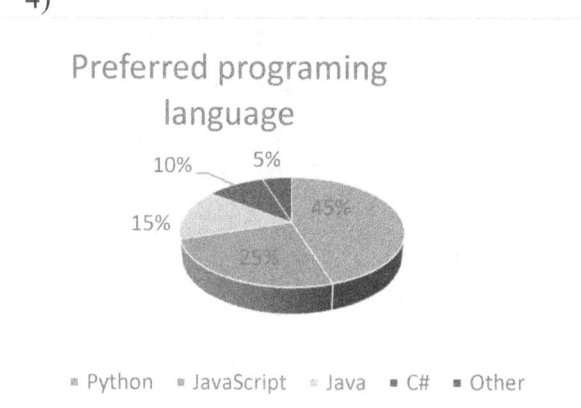

5) ≈ 9%
6) 25%
7) 8,000
8) 2,400
9) ≈ 21%
10) 12 : 7

Quartiles and Outlier
1) $Q1 = 18, Q2 = 25, Q3 = 57$, Outlier=None
2) $Q1 = 100, Q2 = 125, Q3 = 170$, Outlier=None
3) $Q1 = -1, Q2 = 10.5, Q3 = 14$, Outlier=None
4) $Q1 = 2.3, Q2 = 2.855, Q3 = 7.1$, Outlier= 59
5) $Q1 = -0.675, Q2 = 10, Q3 = 29$, Outlier= 87

Box and Whisker Plot
1) $Q1 = 11, Q2 = 12, Q3 = 20$, Min(*lower boundary*) = -2.5 and Max(*upper boundary*) = 33.5

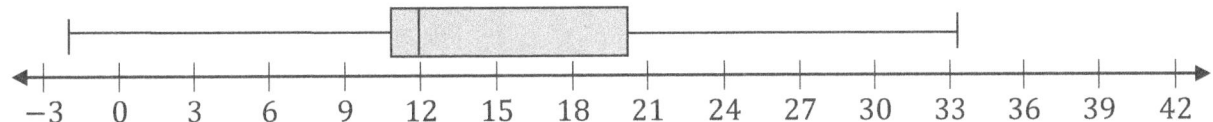

2) $Q1 = 79, Q2 = 86, Q3 = 91$, Min(*lower boundary*) = 61 and Max(*upper boundary*) = 109

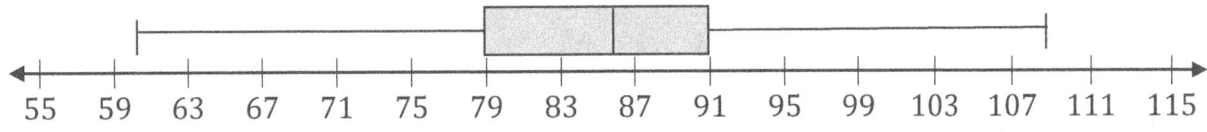

3) $Q1 = 181.5, Q2 = 189, Q3 = 196.5$, Min(*lower boundary*) = 159 and Max(*upper boundary*) = 219

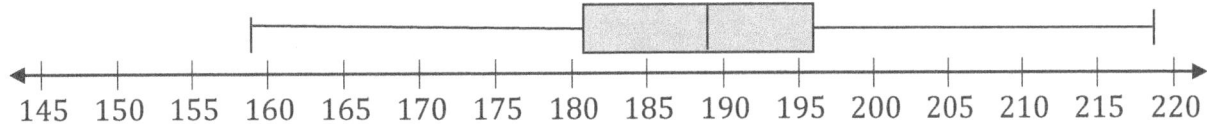

4) $Q1 = 2.1, Q2 = 3.15, Q3 = 3.8$, Min(*lower boundary*) = -0.45 and Max(*upper boundary*) = 6.35

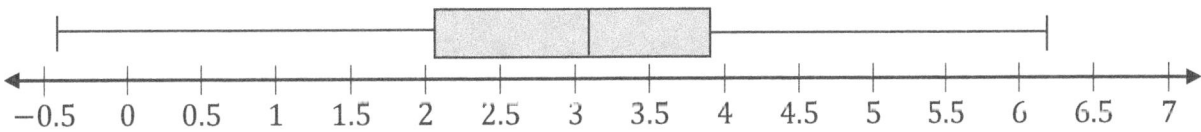

5) $Q1 = -2, Q2 = 6.5, Q3 = 8$, Min(*lower boundary*) = -17 and Max(*upper boundary*) = 23

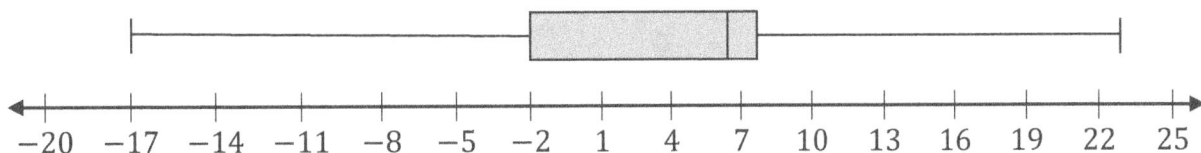

Probability of Simple Events
1) $\frac{3}{8}$
2) $\frac{1}{2}$
3) $\frac{3}{13}$
4) $\frac{4}{15}$
5) $\frac{7}{12}$

Probability of Opposite Events
1) $\frac{5}{6}$
2) $\frac{7}{10}$

3) $\frac{1}{2}$
4) $\frac{34}{75}$

Probability of Mutually Events
1) $\frac{1}{3}$
2) $\frac{8}{13}$

Theoretical Probability
1) $\frac{1}{3}$
2) $\frac{5}{18}$
3) ≈ 0.6

Experimental Probability
1) 0.6
2) 0.6
3) 0.8

Make Predictions
1) 12 heads
2) Approximately 7 or 8 times
3) Approximately 12 wins

Compound Events
1) $\frac{1}{221}$
2) $\frac{1}{6}$
3) 0.18

Probability Word Problems
1) 0.25
2) 0.34
3) 0.64
4) $\frac{4}{5}$
5) $\frac{1}{9}$

5) $\frac{1}{16}$

3) $\frac{1}{4}$
4) 1
5) $\frac{8}{15}$

4) ≈ 0.05
5) $\frac{1}{13}$

4) 0.8
5) $\frac{19}{30}$

4) Experimental $P(Sum > 7) = 0.3$ and Predicted occurrences of $Sum > 8$ in 100 rolls would be $25 - 30$
5) Approximately 600 people

4) 0.49
5) $\frac{6}{95}$

6) 0.25
7) 0.6
8) 0.5
9) 3 yellow marbles
10) 0.75

Chapter 15: Practice Test Review

Ready to put your Grade 7 math skills to the test? Below you'll find **full-length RICAS practice exams** designed to mirror the real testing experience. Use them to identify strengths, spot areas for improvement, and build confidence before the big day.

Before You Begin

- **Gather supplies:** a sharp pencil, eraser, and scratch paper.
- **Take your time:** this rehearsal is untimed, so focus on accuracy and reasoning.
- **Guess if you're unsure:** there's no penalty for wrong answers, so give every question your best shot.
- **No calculators:** Grade 7 RICAS Mathematics must be completed without one.
- **Plan to review:** Once you finish, check your work against the **Practice Test Answer Key** and study any mistakes.

When you're finished, celebrate the progress you've made, each practice test is a step toward stronger skills and greater confidence.

Good luck, and happy problem-solving!

Reference Material

RICAS Mathematics Formula Sheet Grade 7

CONVERSIONS

1 cup = 8 fluid ounces	1 inch = 2.54 centimeters	1 pound = 16 ounces
1 pint = 2 cups	1 meter = 39.37 inches	1 pound = 0.454 kilogram
1 quart = 2 pints	1 mile = 5,280 feet	1 kilogram = 2.2 pounds
1 gallon = 4 quarts	1 mile = 1,760 yards	1 ton = 2,000 pounds
1 gallon = 3.785 liters	1 mile = 1.609 kilometers	
1 liter = 0.2642 gallon	1 kilometer = 0.62 mile	
1 liter = 1,000 cubic centimeters		

AREA (A) FORMULAS

Square…………………. $A = s^2$

Rectangle………………. $A = bh$

OR $A = lw$

Parallelogram …………. $A = bh$

Triangle ………………. $A = \frac{1}{2}bh$

Trapezoid ……………. $A = \frac{1}{2}h(b_1 + b_2)$

CIRCLE FORMULAS

Circle …………………. $A = \pi r^2$

Circle …………………. $C = 2\pi r$

OR $C = \pi d$

VOLUME (V) FORMULAS

Cube ………………… $V = s^3$

(s = length of an edge)

Right prism……………… $V = Bh$

TOTAL SURFACE AREA (SA) FORMULAS

Right Rectangular Prism

$$SA = 2(lw) + 2(hw) + 2(lh)$$

www.mathnotion.com

RICAS Practice Test

1. Evaluate the expression: $12 - 3 \times 4 - |2 - 5|$
   A. -3
   B. 3
   C. 9
   D. 15

2. A submarine is at -250 feet. It rises 180 feet, then dives 70 feet. What is its new depth?
   A. $120\ ft$
   B. $-120\ ft$
   C. $140\ ft$
   D. $-140\ ft$

3. What are the GCF and LCM of 72 and 30?
   A. GCF = 6 and LCM = 72
   B. GCF = 30 and LCM = 360
   C. GCF = 6 and LCM = 360
   D. GCF = 360 and LCM = 6

4. Round 8.4973 to the nearest hundredth.
   A. 8.49
   B. 8.50
   C. 8.4
   D. 8.5

5. A farmer harvested 120.5 kg of apples. They sold 4 bags of 12.75 kg each and used 3.5 kg for baking. How many kg remain?
   A. 64.5 kg
   B. 66 kg
   C. 65.5 kg
   D. 65 kg

6. Convert $\frac{3}{16}$ to decimal.
   A. 0.1875
   B. 3.16
   C. 3.0625
   D. 0.0625

7. Evaluate $2\frac{1}{15} - 1\frac{3}{10} \times \frac{1}{3} + \frac{9}{20}$.
   A. $\frac{17}{20}$
   B. $\frac{1}{20}$
   C. $\frac{25}{12}$
   D. $\frac{41}{20}$

7th Grade Rhode Island Math

8. On average, Simone drinks $\frac{3}{4}$ of a 16-ounce glass of coffee in $\frac{2}{5}$ hour. How much coffee does she drink in an hour?

    A. 45 ounces
    B. 40 ounces
    C. 30 ounces
    D. 15 ounces

9. Arrange from least to greatest: $-1.25, \frac{3}{4}, -\frac{5}{2}, 0.6, -1$

    A. $-\frac{5}{2} < -1.25 < -1 < 0.6 < \frac{3}{4}$
    B. $-\frac{5}{2} < -1.25 < -1 < \frac{3}{4} < 0.6$
    C. $-1.251 < -\frac{5}{2} < -1 < 0.6 < \frac{3}{4}$
    D. $-1 < -1.25 < -\frac{5}{2} < 0.6 < \frac{3}{4}$

10. Solve for x: $\frac{12}{30} = \frac{16}{x}$

    A. 34
    B. 36
    C. 40
    D. 16

11. A printing company has two printers, Printer A and Printer B. Printer A can print 180 pages in 12 minutes. Printer B can print at a rate that is $\frac{3}{4}$ the speed of Printer A. Both printers start working together to print a job of 1,800 pages, but Printer B stops after 20 minutes, while Printer A continues until the job is done. How long (in total minutes) does it take to finish the entire 1,800-page print job?

    A. 85 minutes
    B. 95 minutes
    C. 100 minutes
    D. 105 minutes

12. A $120 jacket is on sale for 25% off. After the discount, a 6% sales tax is applied. What is the final price?

    A. $90.60
    B. $95.40
    C. $100.80
    D. $105.20

13. $5,000 is invested at 6% annual interest compounded yearly for 3 years. What is the total amount?

    A. $5,900
    B. $5,955.08
    C. $6,182.48
    D. $6,370.38

14. $\sqrt{\frac{49x^{11}y^8}{81x^3y^4}}$?

   A. $\frac{7}{9}x^8y^4$

   B. $\frac{7y^2}{9x^4}$

   C. $\frac{7}{9}x^4y^2$

   D. $2\frac{7}{9}x^2y^4$

15. What is 187,000,000 in scientific notations?

   A. 187×10^6

   B. 18.7×10^7

   C. 1.87×10^8

   D. 0.187×10^9

16. Which two integers is $-\sqrt{50} + 2$ between?

   A. -6 and -5

   B. -7 and -8

   C. 6 and 5

   D. 7 and 8

17. Which algebraic expression represents "5 less than twice a number x"?

   A. $5 - 2x$

   B. $2x - 5$

   C. $5x - 2$

   D. $2 - 5x$

18. If $x = -2$ and $y = 3$, evaluate $2x^2 - xy + 4y$

   A. 10

   B. 14

   C. 22

   D. 26

19. A shop sells a product for $120 each. The cost to produce each item is $80, and the fixed monthly cost is $2000. What is the profit P if n items sold in a month?

   A. $P = 40n - 2000$

   B. $P = 120n + 2000$

   C. $200n - 80$

   D. $2000 - 40n$

20. Factor completely: $12x^3y^2 - 6yx^2 + 18x^4y^3$

   A. $6x^2y(2xy + 3x^2y^2)$

   B. $2x^3y^2(6 - 3xy + 9xy)$

   C. $6x^2y(2xy - 1 + 3x^2y^2)$

   D. $3xy^2(4x^2 - 2y + 6x^3y)$

21. Which equation represents the line with slope -2 and y-intercept 4 in slope-intercept form?

    A. $2x + y = 4$
    B. $y = -2x + 4$
    C. $y = 4x - 2$
    D. $y = -2x - 4$

22. Solve for y: $2(y - 4) + 5 = 15 - 2y$

    A. 4
    B. 3
    C. -13
    D. $\frac{9}{2}$

23. Which number line shows the solution to the inequality $-4x + 7 < -5$?

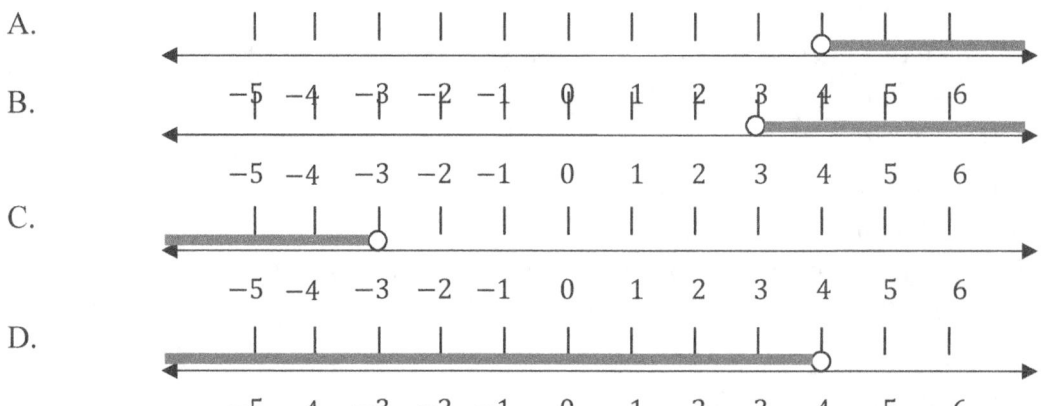

A.
B.
C.
D.

24. To join a gym, there is a $50 fee and a $30 monthly charge. How many months can you join if you have at most $200?

    A. $m \leq 5$
    B. $m \geq 5$
    C. $m \leq 6$
    D. $m \geq 6$

25. Identify the slope and y-intercept of the line: $3x - 4y = 12$

    A. Slope= 3, y-intercept= 12
    B. Slope= $\frac{3}{4}$, y-intercept= -3
    C. Slope= $-\frac{3}{4}$, y-intercept= 3
    D. Slope= $\frac{4}{3}$, y-intercept= -4

26. Write the equation of the line passing through the two points $(1, -3)$ and $(4, 2)$.

    A. $3y + 5x = 14$
    B. $5y + 3x = -6$
    C. $3y - 5x = -14$
    D. $5y - 3x = 6$

27. Which equation is the equation for the following line?

   A. $y = -\frac{3}{4}x + 1$
   B. $y = x$
   C. $y = 4x - 1$
   D. $y = \frac{3}{4}x + 1$

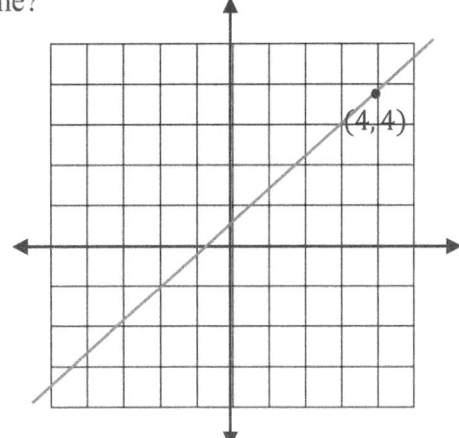

28. Solve the system: $\begin{cases} 2x + y = 5 \\ x - y = 1 \end{cases}$

   A. $(2, 1)$
   B. $(1, 2)$
   C. $(3, -1)$
   D. $(-1, 3)$

29. What is the image of point $B(4,7)$ when reflected over the x-axis?

   A. $(4, -7)$
   B. $(-4, 7)$
   C. $(-4, -7)$
   D. $(7, 4)$

30. A model car is a $\frac{1}{20}$ scale of the real car. If the model's length is 9 inches, what is the real car's length?

   A. 18 feet
   B. 20 feet
   C. 22 feet
   D. 15 feet

31. A map uses a scale of 1 cm ∶ 5 km. If two cities are 3.5 cm apart on the map, what is the actual distance?

   A. 15.5 km
   B. 17.5 km
   C. 20 km
   D. 22.5 km

32. Which arithmetic sequence is represented by the expression $6m - 1$, where m represents the position of s term in sequence?

   A. 11, 17, 23, 28, 32, ...
   B. 5, 11, 17, 23, 29, ...
   C. 5, 11, 15, 22, 27, ...
   D. 8, 11, 16, 19, 25, ...

www.mathnotion.com

33. A car depreciates 15% annually. If its initial value is $20,000, what is its value after 3 years?
    A. ≈ $15,700.00
    B. ≈ $14,450.00
    C. ≈≈ $12,282.50
    D. ≈ $16,500.00

34. Two similar hexagons have side lengths in the ratio 3: 5. What is the ratio of their areas?
    A. 3: 5
    B. 6: 10
    C. 27: 25
    D. 9: 25

35. Which condition proves △ ABC ≅△ DEF?
    A. $AB = DE$, $BC = EF$, $\angle B = \angle E$
    B. $AB = DE$, $AC = DF$
    C. $\angle A = \angle D$, $\angle B = \angle E$, $\angle C = \angle F$
    D. All of the above

36. Line P, R and S intersect each other, as shown in the diagram below. Based on the angle measures, what is the value of θ?
    A. 120°
    B. 132°
    C. 122°
    D. 102°

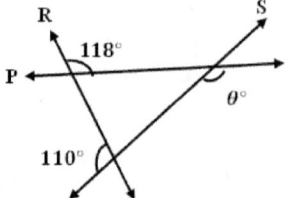

37. A ladder is leaning against a vertical wall. The foot of the ladder is 6 feet away from the base of the wall. The ladder reaches a point on the wall that is 8 feet below the top of the 26-foot wall. How long is the ladder?
    A. $6\sqrt{10}$ feet
    B. $8\sqrt{10}$ feet
    C. 18 feet
    D. 8 feet

38. The circumference of a circle is 14 π centimeters. What is the area of the circle in terms of π?
    A. 14π
    B. 149π
    C. 49π
    D. 98π

39. Find the perimeter of the following figure:
    A. $28 + 2\pi$ m
    B. $28 + 4\pi$ m
    C. $24 + 2\pi$ m
    D. $24 + 4\pi$ m

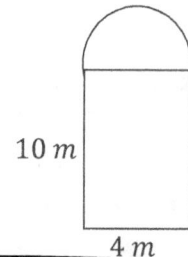

40. A storage box in the shape of a rectangular prism has a length that is 3 times its width, and its height is 4 inches more than its width. If the volume of the box is 1,080 cubic inches, what is the width of this box?

    A. 7 in
    B. 6 in
    C. 5 in
    D. 4 in

41. A group of employees have their weight recorded to make a data set. The mean, median, mode, and range of the data set are recorded. Then, the weight of the manager is included in making a new data set. The manager's weight is more than all but one of the employees. Which measure must be the same when the manager's weight is included?

    A. Mean
    B. Median
    C. Mode
    D. Range

42. Which statement is true about the histogram below?

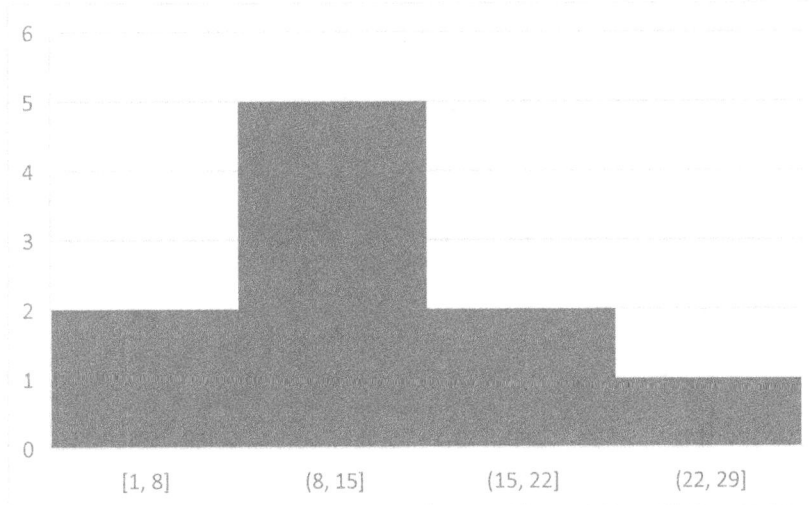

    A. Most data fall in 15 − 22
    B. The range is 20
    C. The mode is in 8 − 15
    D. The median is in 22 − 29

43. In a box plot $Q_1 = 60, Q_3 = 80$ and median $= 70$. What is the interquartile range (IQR)?

    A. 10
    B. 15
    C. 20
    D. 30

44. A coin flipped 50 times, landing on heads 27 times. What is the experimental probability of tails?

   A. $\frac{23}{50}$
   B. $\frac{27}{50}$
   C. $\frac{1}{2}$
   D. $\frac{23}{27}$

45. A store has 20 laptops, 5 of which are defective. If 2 are randomly chosen, what is the probability both are defective?

   A. $\frac{1}{20}$
   B. $\frac{1}{19}$
   C. $\frac{1}{10}$
   D. $\frac{1}{5}$

STOP

"This is the end of the test. You may check your work if you still have time."

Answer Key

Now it's time to check your answers! Review your results to understand any mistakes and find out which areas you can improve on.

Practice Test key

#	Ans	#	Ans	#	Ans
1	A	16	A	31	B
2	D	17	B	32	B
3	C	18	D	33	C
4	B	19	A	34	D
5	B	20	C	35	A
6	A	21	B	34	B
7	C	22	D	37	A
8	C	23	B	38	C
9	A	24	A	39	C
10	C	25	B	40	B
11	D	26	C	41	D
12	B	27	D	42	C
13	B	28	A	43	C
14	C	29	A	44	A
15	C	30	D	45	B

www.mathnotion.com

Answers and Explanations

1) **Order of Operations: Answer: A**
   According to acronym PEMDAS we have:
   $12 - 3 \times 4 - |2 - 5| = 12 - 12 - |2 - 5| = 0 - (3) = -3$

2) **Integer Numbers: Answer: D**
   New depth $= -250 + 180 - 70 = -140 \, ft$

3) **GCF and LCM: Answer: C**
   $72 = 2^3 \times 3^2$ and $30 = 2^1 \times 3^1 \times 5^1$.
   For finding GCF take the lowest powers of the common prime factors: GCF$= 2^1 \times 3^1 = 6$.
   For finding LCM take the highest powers of all primes that appear: LCM $= 2^3 \times 3^2 \times 5 = 8 \times 9 \times 5 = 360$

4) **Rounding Decimals: Answer: B**
   Hundredths place is 9. The thousandths digit (7) $\geq 5 \to$ round up. $8.4973 \to 8.50$

5) **Operation on Decimals: Answer: B**
   Total sold: $4 \times 12.75 = 51$ kg.
   Total used/sold: $51 + 3.5 = 54.5$ kg. Remaining: $120.5 - 54.5 = 66$ kg.

6) **Converting Fractions to Decimals: Answer: A**
   To convert the fraction $\frac{3}{16}$ to decimal, we divide the numerator (3) by the denominator (16):
   $3 \div 16 = 0.1875$

7) **Operations on Fractions: Answer: C**
   According to the order of operations, first we do the multiplication: $1\frac{3}{10} \times \frac{1}{3} = \frac{13}{10} \times \frac{1}{3} = \frac{13}{30}$.
   Now we have: $2\frac{1}{15} - 1\frac{3}{10} \times \frac{1}{3} + \frac{9}{20} = \frac{31}{15} - \frac{13}{30} + \frac{9}{20}$.
   The least common denominator of $15, 30,$ and 20 is 60:
   $\frac{31}{15} - \frac{13}{30} + \frac{9}{20} = \frac{124}{60} - \frac{26}{60} + \frac{27}{60} = \frac{125}{60} = \frac{25}{12}$

8) **Word Problem: Answer: C**
   Find how many ounces she drinks in $\frac{2}{5}$ hour: $\frac{3}{4} \times 16 = 12$ ounces.
   To find the rate per hour, divide by $\frac{2}{5}$: $\frac{12}{\frac{2}{5}}$ $(12 \div \frac{2}{5}) = 12 \times \frac{5}{2} = 30$ ounces.

9) **Ordering Rational Numbers: Answer: A**
   Convert all to decimal: $-\frac{5}{2} = -2.5$ and $\frac{3}{4} = 0.75$. Order: $-2.5 < -1.25 < -1 < 0.6 < 0.75$

10) **Solving Proportions: Answer: C**
    Picture it like scaling a recipe: if 12 "whatevers" match up with 30 "thingies," how many "whatevers" go with 100 "thingies?" Keep the ratio:
    $\frac{12}{30} = \frac{x}{100}$. Cross-multiply—$12 \times 100 = 30x$—and you get $x = 40$.

www.mathnotion.com

11) **Rate and Ratios: Answer:** D

Printer A: It prints 180 pages in 12 minutes, so its rate is: $\frac{180}{12} = 15$ pages per minute.

Printer B: It prints at $\frac{3}{4}$ the speed of Printer A. Rate is: $\frac{3}{4} \times 15 = 11.25$ pages per minute.

Both printers work together for 20 minutes, so:

Printer A prints: $15 \times 20 = 300$ pages

Printer B prints: $11.25 \times 20 = 225$ pages

The total pages printed first 20 minutes: $300 + 225 = 525$ pages.

Pages remaining: $1800 - 525 = 1{,}275$ pages.

Printer A prints 15 pages per minute, so: $\frac{1{,}275}{15} = 85$ minutes.

Total time to finish the job: $20 + 85 = 105$ minutes.

12) **Discount and Tax: Answer:** B

Discount: $120 \times 0.25 = 30 \rightarrow$ New price $= 120 - 30 = 90$. Tax: $90 \times 0.06 = 5.4$
$\rightarrow$ Final price $= 90 + 5.40 = 95.40$.

13) **Compound Interest: Answer:** B

Compound interest formula: $A = P\left(1 + \frac{r}{n}\right)^{nt}$.

Here, $n = 1$, so $A = 500 \times (1.06)^3 \approx 5000 \times 1.191016 \approx 5955.08$

14) **Exponent and Square Root: Answer:** C

$\sqrt{\frac{49x^{11}y^8}{81x^3y^4}} = \sqrt{\frac{49}{81}} \times \sqrt{\frac{x^{11}}{x^3}} \times \sqrt{\frac{y^8}{y^4}} = \frac{\sqrt{49}}{\sqrt{81}} \times \sqrt{x^{11-3}} \times \sqrt{y^{8-4}} = \frac{7}{9}\sqrt{x^8} \times \sqrt{y^4} = \frac{7}{9}x^4y^2$

15) **Scientific Notations: Answer:** C

Move the decimal point 8 places to the left to get a number between 1 and 10:
$187{,}000{,}000 = 1.87 \times 10^8$

16) **Estimating Square Roots: Answer:** A

Since $\sqrt{49} < \sqrt{50} < \sqrt{64}$, we have $7 < \sqrt{50} < 8$. Negating reverses the inequality, giving: $-8 > -\sqrt{50} > -7$. Adding 2 throughout yields $-8 + 2 > -\sqrt{50} + 2 > -7 + 2$. Therefore, $-6 > -\sqrt{50} + 2 > -5$ Therefore, $-\sqrt{50} + 2$ lies between -6 and -5.

17) **Translating Phrases: Answer:** B

"Twice a number x" is $2x$, and "5 less than" means subtract 5.

18) **Evaluating Expressions: Answer:** D

$2(-2)^2 - (-2)(3) + 4(3) = 2(4) + 6 + 12 = 8 + 6 + 12 = 26$.

19) **Algebraic Expressions: Answer:** A

Profit=Revenue−Cost: $P = 120n - (80n + 2000) = 40n - 2000$.

20) **Factoring Using Distributive Property: Answer:** C

GCF of coefficients $= 6$; GCF of variables $= x^2y$.

Factor out $6x^2y$: $6x^2y(2xy - 1 + 3x^2y^2)$

21) **Forms of Linear Equations: Answer:** B

Slope-intercept form is $y = mx + b$. Given $m = -2$ and $b = 4$: $y = -2x + 4$

22) **Multi Step Equation: Answer:** D

$2(y-4) + 5 = 15 - 2y \rightarrow 2y - 8 + 5 = 15 - 2y \rightarrow 2y + 2y = 15 + 3$
$\rightarrow 4y = 18 \rightarrow y = \frac{18}{4} = \frac{9}{2}$

23) **Graphing Inequalities: Answer:** B

$-4x + 7 < -5 \rightarrow -4x < -5 - 7 \rightarrow -4x < -12 \rightarrow x > -\frac{12}{-4} \rightarrow x > 3$

24) **Inequality: Answer:** A

Total cost: $50 + 30m \leq 200 \rightarrow 30m \leq 150 \rightarrow m \leq \frac{150}{30} \rightarrow m \leq 5$.

25) **Slope and Y-Intercept from Equation: Answer:** B

Rewrite in slope-intercept form $y = mx + b$: $-4y = -3x + 12 \rightarrow y = \frac{3}{4}x - 3$.
Slope $= \frac{3}{4}$, y-intercept $= -3$

26) **Writing Linear Equation: Answer:** C

Slop $= \frac{2-(-3)}{4-1} = \frac{5}{3}$. The point-slope form of a line is: $y - y_1 = m(x - x_1)$.
Use the point $(4, 2)$ and the slope $\frac{5}{3}$: $y - 2 = \frac{5}{3}(x - 4) \rightarrow y - 2 = \frac{5}{3}x - \frac{20}{3}$
$y = \frac{5}{3}x - \frac{20}{3} + 2 \rightarrow y = \frac{5}{3}x - \frac{14}{3} \rightarrow 3y = 5x - 14 \rightarrow 3y - 5x = -14$

27) **Graphing a Line: Answer:** D

Using two points $4, 4)$ and $(0, 1)$ from graph and point-slope form we have:
Slope $= \frac{4-1}{4-0} = \frac{3}{4} \rightarrow y - 1 = \frac{3}{4}(x - 0)$, so $y = \frac{3}{4}x + 1$

28) **System of Equations: Answer:** A

Add the equations: $3x = 6 \rightarrow x = 2$.
Substitute into the second equation: $2 - y = 1 \rightarrow y = 1$.

29) **Reflections: Answer:** A

Reflection over x-axis changes the sign of the y-coordinate: $(4, 7) \rightarrow (4, -7)$

30) **Scale Drawings: Answer:** D

Real length = Model length × Scale factor → $9 \times 20 = 180$ inches = 15 feet.

31) **Dilations: Answer:** B

Actual distance $= 3.5 \times 5 = 17.5$ km

32) **Arithmetic Sequence: Answer:** B

$m = 1: 6(1) - 1 = 5,$ $m = 4: 6(4) - 1 = 23,$
$m = 2: 6(2) - 1 = 11,$ $m = 5: 6(5) - 1 = 29.$
$m = 3: 6(3) - 1 = 17,$

33) **Geometric Sequence: Answer:** C

Geometric decay formula: $a_n = a_1 \times (1 - r)^n$. Here $a_1 = 20{,}000, r = 0.15, n = 3$:
$a_3 = 20{,}000 \times (0.85)^3 \approx 20{,}000 \times 0.6141 \approx 12{,}282.50$

34) **Ratio of Areas in Similar Figures: Answer:** D

Area ratio = square of side ratio = $(3:5)^2 = 9:25$.

35) **Congruence Criteria for Triangles: Answer: A**

A: SAS (Side-Angle-Side), and B and C does not guarantee congruence.

36) **Angles in Triangles: Answer: B**

$\hat{x} = 180° - 118° = 62°, \hat{y} = 180° - 110° = 70°$, Since angles
$x, y,$ and z are the angles of a triangle, their sum is $180°$ degrees:
$\hat{x} + \hat{y} + \hat{z} = 180° \rightarrow 62° + 70° + \hat{z} = 180°$
$\rightarrow \hat{z} = 180° - 62° - 70° = 48° \rightarrow \hat{z} + \hat{\theta} = 180°$
$\rightarrow 48° + \hat{\theta} = 180° \rightarrow \hat{\theta} = 180° - 48° = 132°$

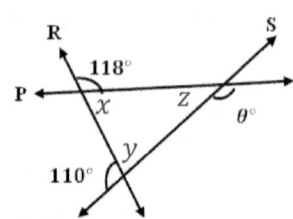

37) **Pythagorean Theorem: Answer: A**

The ladder forms the hypotenuse of a right triangle, and it reaches a point 8 feet below the top, so the height where the ladder touches $= 26 - 8 = 18$ feet. We now have a right triangle with one leg $= 6$ ft and the other leg $= 18$ ft. using Pythagorean Theorem:
$a^2 + b^2 = c^2 \rightarrow 6^2 + 18^2 = c^2 \rightarrow c^2 = 360 \rightarrow c = \sqrt{360} = 6\sqrt{10}$

38) **Area of Circles: Answer: C**

Circumference $= 2\pi r = 14\pi \rightarrow 2r = 14 \rightarrow r = 7$. Area $= \pi r^2 = \pi 7^2 = 49\pi$

39) **Compound Figures: Answer: C**

Rectangle perimeter (without top) $= 10 + 4 + 10 = 24\ m$.
Semicircle circumference $= \pi(2) = 2\pi\ m$. Total perimeter $= 24 + 2\pi\ m$.

40) **Volume of Rectangular Prism: Answer: B**

Let the width be x inches, then: Length $= 3x$ and Height $= x + 4$. By trying all options and using the volume formula: Volume= length×width×height,
we will find if $x = 6$, then length $= 18$ and height $= 10 \Rightarrow$ Volume= $6 \times 18 \times 10 = 080\ in^3$

41) **Mean, Median, Mode and Range: Answer: D**

Range = Maximum−Minimum, the manager's weight is less than the highest, so the maximum doesn't change. Since the manager is not the lightest, the minimum also doesn't change. So, the range will stay the same. Mean, median and mode my change.

42) **Histograms: Answer: C**

Mode is the most frequent value/bin.
$8 - 15$ has the highest frequency (5), so it's the modal class.

43) **Box and Whisker Plot: Answer: C**

IQR= $Q_3 - Q_1 = 80 - 60 = 20$

44) **Experimental Probability: Answer: A**

Tail $= 50 - 27 = 23$. Experimental P(Tail)=$\frac{23}{50}$

45) **Probability: Answer: B**

P (1st defective) $= \frac{5}{20} = \frac{1}{4}$, P (2nd defective) $= \frac{4}{19}$ then P (Both defective) $= \frac{1}{4} \times \frac{4}{19} = \frac{1}{19}$

"End"

www.mathnotion.com

www.ingramcontent.com/pod-product-compliance
Lightning Source LLC
Chambersburg PA
CBHW082206070526
44585CB00020B/2309